MAYA SALT WORKS

Maya Studies

UNIVERSITY PRESS OF FLORIDA

Florida A&M University, Tallahassee
Florida Atlantic University, Boca Raton
Florida Gulf Coast University, Ft. Myers
Florida International University, Miami
Florida State University, Tallahassee
New College of Florida, Sarasota
University of Central Florida, Orlando
University of Florida, Gainesville
University of North Florida, Jacksonville
University of South Florida, Tampa
University of West Florida, Pensacola

MAYA SALT WORKS

HEATHER MCKILLOP

Foreword by Arlen F. Chase and Diane Z. Chase

University Press of Florida
Gainesville · Tallahassee · Tampa · Boca Raton
Pensacola · Orlando · Miami · Jacksonville · Ft. Myers · Sarasota

Published in the United States of America. Printed on acid-free paper.

This book may be available in an electronic edition.

24 23 22 21 20 19 6 5 4 3 2 1

Library of Congress Cataloging-in-Publication Data
Names: McKillop, Heather Irene, author.
Title: Maya salt works / Heather McKillop.
Other titles: Maya studies.
Description: Gainesville : University Press of Florida, 2019. | Series: Maya studies | Includes bibliographical references and index.
Identifiers: LCCN 2018046945 | ISBN 9780813056333 (cloth : alk. paper)
Subjects: LCSH: Salt industry and trade—Belize. | Mayas—Commerce. | Mayas—Belize—Antiquities. | Salt—Belize—History.
Classification: LCC F1435.3.S24 M35 2019 | DDC 972.82/01—dc23
LC record available at https://lccn.loc.gov/2018046945

The University Press of Florida is the scholarly publishing agency for the State University System of Florida, comprising Florida A&M University, Florida Atlantic University, Florida Gulf Coast University, Florida International University, Florida State University, New College of Florida, University of Central Florida, University of Florida, University of North Florida, University of South Florida, and University of West Florida.

University Press of Florida
2046 NE Waldo Road
Suite 2100
Gainesville, FL 32609
http://upress.ufl.edu

CONTENTS

FIGURES

TABLES

FOREWORD

Salt is a desired commodity for ancient and contemporary human populations worldwide. More than a condiment, salt is also an excellent food preservative, allowing for both storage and transport of perishable foodstuffs. It is both a luxury for and a mainstay of human civilization.

Heather McKillop has devoted a substantial portion of her career to understanding the ways in which the ancient Maya produced salt and how it was distributed both locally and to other ancient Maya populations. This work constitutes her second book on the subject. Her original tome, *Salt: White Gold of the Maya*, was published by the University Press of Florida in 2002 as the first volume in our Maya Studies series. Her work established the southern coast of Belize as an area of salt production that contributed substantially to the already known production areas on the northern coast of the Yucatán Peninsula and in the Guatemalan highlands. Almost two decades have passed since the appearance of her first book and this one. During this time Heather has continued to investigate the production of ancient Maya salt along the coast of southern Belize and has made a series of new discoveries yielding information extremely relevant to our knowledge about ancient Maya economies.

Much of the work that she has undertaken was literally underwater. Because of sea-level rise, many of the ancient Maya salt production locations have been submerged in shallow lagoons in southern Belize. However, these waterlogged sites have also seen unusual preservation of wood, especially ancient posts that once formed buildings and fences or breaks. The careful recording of these posts has permitted a reconstruction of ancient buildings and activity locales for ancient Maya salt making along the Belizean coast. This work suggests that the Maya were aware of changing sea levels and adapted accordingly by moving further away from the water's edge as it gradually rose. Careful recording of associated artifactual debris that is clustered near the underwater wooden features shows that production took place year-round and has provided some idea of the time depth of salt making in this part of the world. These artifactual materials also reveal trade connections to other parts of the Maya area, suggesting where the coastal Belizean salt may have ultimately been distributed. Much of the salt production that has been found in southern Belize dates to the height of the ancient Maya civilization between AD 650 and 900, and the provisioning of between 480 and 4,800 tons of salt per year by the producers in southern Belize was sufficient to supply this staple to over one million people in the Southern Maya Lowlands. However, even though salt may have been a highly desired commodity that could be transported long distances by sea in canoes, salt production along the coast of southern Belize came to a crashing halt as demand declined at the time of the Maya collapse in the Southern Lowlands.

This book presents important new information on Maya salt production and demonstrates how our thinking about the Maya economies of the Classic period has changed over the last twenty years. The material in this and the earlier salt volume provides a solid foundation for current and future researchers who are interested in the role that salt played in ancient economies.

Arlen F. Chase and Diane Z. Chase
Series Editors

ACKNOWLEDGMENTS

I am grateful for financial support from the following sources to carry out the research discussed in this book: A Faculty Research Grant from Louisiana State University supported the fieldwork in 2004 including the discovery of the K'ak' Naab' Maya wooden canoe paddle and the wooden architecture preserved below the sea floor. A grant from the Foundation for the Advancement of Mesoamerican Studies supported mapping of the K'ak' Naab' site in 2005. National Science Foundation grant #0513398, "Mapping Ancient Maya Wooden Architecture on the Sea Floor, Belize," supported field and lab research from 2006 to 2008. In addition, REU (Research Experiences for Undergraduates) Supplements supported undergraduates in the field in 2006 and 2008. A grant from the National Geographic Society funded field research in 2006. National Science Foundation collaborative award 1331428 (Heather McKillop, Karen McKee, Harry Roberts, and Terance Winemiller) funded the excavations and the environmental and 3D GIS research referenced in this book, which focuses on the discovery and mapping. A Louisiana Board of Regents grant funded equipment to start the LSU Digital Imaging and Visualization in Archaeology grant in 2009, which was used to make 3D digital images and 3D prints of artifacts and wooden architecture on the underwater Maya mapping project. A Site Preservation Grant from the Archaeological Institute of America provided funds for the exhibits in Punta Gorda and the Paynes Creek Ranger Station featuring 3D-printed replicas of artifacts and posts from the fieldwork.

The field research was carried out with annual permits, friendship, and encouragement from the Belize government's Institute of Archaeology, notably of former director Dr. Jaime Awe, current director Dr. John Morris, and Institute staff including Melissa Badillo, Antonio Beardall, and others. Our host family at Village Farm, Tanya Russ and John Spang, provided an inspiring research location for my team to be based as well as insights, good conversation, and encouragement, for which I am very grateful. I appreciate the rangers at the ranger stations in Paynes Creek National Park and Port Honduras Marine Reserve from the Toledo Institute for Development and the Environment (TIDE) as well as the executive director of TIDE, Celia Mahung, and her staff for their conservation efforts. I am grateful to the Toledo Tour Guide Association and the Toledo Branch of the Belize Tourism Association Industry for encouraging the fieldwork and providing a venue for the exhibits featuring 3D-printed replicas of Paynes Creek underwater Maya artifacts at the Toledo Tourism Association building in Punta Gorda and the Paynes Creek Ranger Station. Many friends and businesses in Punta Gorda, the nearest town to the field project, encouraged our research, and I appreciate their friendship and interest in the archaeology, notably Paul Mahung of LoveFM, Iris and Alejandro Vernon, Bobbie Polonio, Irene and Peter Mahung, Dwayne and MiLady Wagner, Josephine Supaul, Leonor Requena, Wallace Young and family, Dorna Young, Inez Coc, Dr. Maria Luz Legra Williams, Tina and Kirby Salisbury, Dennis Garbutt, Amber and Paul Carpenter, and Chet Schmidt and family.

The field research was made possible by enthusiasm, hard work, and flexibility of the Louisiana State University graduate and undergraduate students, including Cory Sills, Bretton Somers, Mark Robinson, Zoe Morris, Amanda Evans, Rachel Watson, Roberto Rosado, Michael Mirobelli, Kevin Pemberton, Amanda Pitcock, Jessica Harrison, Taylor Aucoin, Jaclyn Landry, and Matt Helmer. John Young was not only project boat driver but a member of the field team from 2004 throughout the project, and I appreciate his enthusiasm, boat and field skills, and friendship. In the lab, undergraduate student workers assisted in artifact conservation (changing water in containers of wet wood and wet pottery samples), notably Amanda Pitcock, Jessica Harrison, David Susko, Patrick Vines, Tajii Schmidt, Abigail Woltering, and Jaclyn Landry. Students assisted with 3D scanning, sorting artifact photos and drawings, entering data and field notes, compiling digital inventories, and related lab work in the DIVA Lab, notably Bretton Somers, Michael Mirobelli, Cory Sills, Valerie Feathers, Victoria Harrington, Kelsey Johnson, Kurt Dilores, Jane Fiegel, Amelia Juneau, Cher Foster, April Belleau, Christina Brown, Emily Cook, Patrick Vines, David Susko, Abigail Woltering, Katherine Shurik, Karen Bebee, Travis Munson, and Kobi Weaver. I appreciate the work of Mark

Robinson in identifying wood in the archaeology lab and his supervision of students in the lab as well. Long days in the lagoon and in the lab were spent with these smart students discussing the archaeology of salt.

Many people have advanced my ideas on the economics of ancient Maya salt. For broadening my horizons beyond the Maya, I am grateful to Lothar von Falkenhausen and Shuicheng Li for including me in international salt workshops in Germany and China, and to the colleagues I met, notably Rowan Flad, Gwen Bennet, Pochan Chan, Laurent Olivier, Oliver Weller, Thomas Saile, Martin Hees, Ian Brown, Hans Ulrich Vogel, and others. Other archaeology colleagues who have influenced my thinking in discussions include Andrea Yankowski, Ashley Dumas, Paul Eubanks, Eduardo Williams, Tony Andrews, Elizabeth Graham, Robin Robertson, Jeremy Sabloff, Thomas Guderjan, and Cynthia Robin. I am grateful to C. Wayne Smith and Helen DeWolf for conservation of waterlogged wooden artifacts and their advice.

I appreciate that Eleanor, whom I called from the end of the dock at Village Farm during long field seasons from when she was age five, accepted the importance of her mother as an academic and a mother. I appreciate that my husband, Robert Tague Jr., encouraged my field research and other academic pursuits and that he shared fully in raising our daughter.

1

Economics of Maya Salt

The Classic Maya acquired salt by solar evaporation and by boiling brine in pots over fires inside specialized salt kitchens. The brine-boiling method was used at the Paynes Creek Salt Works, on the coast of southern Belize (Figure 1.1). The salt works were submerged by sea-level rise, preserving wooden buildings below the sea floor in red mangrove peat (*Rhizophora mangle*) along with associated artifacts. In addition to briquetage—the pottery used in brine boiling—nonlocal pottery at the salt works can be used to identify the inland consumers of salt. Analyses of the wooden buildings, briquetage, and other artifacts are used to evaluate the scale, intensity, and organization of salt production and its role in the Classic Maya economy.

Other researchers have discovered that Classic Maya householders were not as self-sufficient as once thought but instead produced surplus goods for marketplace trade. In addition, householders obtained goods and resources, including exotics such as obsidian and jadeite, from marketplaces nearby and at more distance locations (Chase et al. 2015; Hirth 2009a; Masson et al. 2016; Sheets et al. 2015). At Ceren—a Classic Maya village that was dramatically preserved by tephra from a volcanic eruption—householders specialized in agricultural or craft products, taking them to markets outside the village where they traded for a variety of goods, including exotics (Sheets et al. 2015). Neighborhood marketplaces at the end of causeways radiating from central Caracol, Belize, provided a location for goods and resources from nearby and more distance locations to be traded (Chase et al. 2015). Objects made from exotic material, such as obsidian blades and jadeite beads, as well as nonlocal painted serving dishes were widely distributed at commoner households throughout the city. The highest quality and quantity of highly crafted goods, often of exotic material, were found in dynastic and other elite contexts at Caracol,

Figure 1.1. The Maya area showing the Paynes Creek Salt Works and other sites mentioned in the text. By Mary Lee Eggart.

but the exotic and other nonlocal materials were still available to most people. Was salt distributed through marketplace trade, along with other goods? In particular, was salt production for home use or surplus household production for marketplace trade? How much salt was produced and to what degree was production state controlled?

Gifting, Tribute, and Feasting

During the Classic period, the ancient Maya reached the height of their civilization. Some 80 city-states were flourishing in the tropical rainforests and highland areas of Guatemala, Belize, the Yucatán of Mexico, western Honduras, and El Salvador (Figure 1.1; Martin and Grube 2000). Each city-state was ruled by a dynastic king or queen with an entourage of royal family members, artisans, courtiers, and others living at the royal court. In addition, the large cities had subsidiary towns and villages in the surrounding countryside.

In contrast to other ancient civilizations that had strong central governments—such as Imperial China, the Aztecs, and the Incas—the Classic Maya lacked a well-developed bureaucracy, an obsession with tax records, or a strong military force. By comparison, they were less centrally organized.[1] Certainly warfare was common, even endemic, according to some researchers, escalating during the Late Classic (Demarest 2004). Initially, the focus by Classic Maya dynastic leaders was on self-aggrandizing, but by the Late Classic there was an interest in control of land and labor.

Royal gift exchange, tribute, and feasting were important means for the dynastic Maya to establish, maintain, and reinforce power relationships (Foias 2007). At events held in the royal courts, Maya rulers brokered and solidified power hierarchically with dynastic leaders of other city-states and asserted their superior status vis-à-vis elite rulers of smaller communities within their polity. The dynastic leaders displayed their power through monumental architecture, stone or stucco depictions of rulers on building facades, and propaganda texts on stelae recounting, sometimes in exaggerated form, their exploits and those of their dynastic forebears. Highly crafted goods, often with value added by their nonlocal origin, were given as gifts, such as the Buenavista vase from Naranjo (Ball and Taschek 2001; Reents-Budet et al. 2000). Some gifts were transported great distances, including the Altun Ha vase found at Copan or jadeite funeral masks on deceased royalty at Palenque and Calakmul.

Classic Maya leaders conquered nearby and distant cities, placing their leaders as rulers and exacting tribute. Their palaces functioned as administrative offices as well as royal residences (Christie 2003; Inomata and Houston 2001). There were court scribes and a hierarchy of elites at the royal courts and in subsidiary communities within each polity, who served in administrative functions and were accountable to the dynastic leaders. Scribes were high-status individuals living in the royal court, as described for Aguateca (Inomata 2001). The written records they have left behind on stone monuments, pottery vessels, and other objects focus on dynastic history with little information on taxation, tribute, or other economic matters. Such economic records may have been written in paper books that did not preserve.

The dynastic Maya exercised their political power through tribute from conquered city-states and subsidiary populations within their realm, in addition to maintaining marketplaces (King 2015). With urban populations reaching 100,000 at the capital city of Tikal (Harrison 1999) and perhaps 110,000 at Caracol (Chase and Chase 2014), agricultural products and labor were in high demand for public works such as the rebuilding and renovation of stone temples, palaces, and causeways. Representations of tribute payment can be seen in painted scenes on pottery vessels and on the temple murals at Bonampak (Houston et al. 2006; Miller 1986). Xuenkal women specialized in cloth production in the Terminal Classic, at a time when the nearby city of Chichén Itzá reached its peak and demanded tribute (Ardren et al. 2016). Textile tribute at Tikal, Calakmul, and other Late Classic cities in the Southern Maya Lowlands expanded as standard sizes of cotton cloth acquired as tribute were used by Maya royalty as stored wealth, used for acquiring goods and food later (Baron 2018a:110), or as tribute ransom for war captives (McAnany 2010:278–281).

In addition to high-quality cotton cloth, other common tribute items were cacao, jadeite, *Spondylus* shell, and quetzal feathers (Baron 2018b), which "formed an archetypal set of items that royal courts were expected to possess in abundance. They are emblematic of successful feasting, luxurious clothing, and symbolically charged personal adornment" (McAnany 2010:291). A large number of artisans and others living in Classic Maya cities had to be fed. Without evidence of extensive storage facilities (in contrast to the Inca warehouses), stored tribute wealth by royalty and urban marketplaces may have served as places where people acquired food as well as goods and resources from elsewhere (Eppich and Freidel 2015; McAnany 2010).

Craft Production

Studies of ancient economies in complex societies, such as the Classic Maya, include factors such as household versus workshop production, independent versus attached workshops, production versus specialization, and multicrafting. Household production focuses on extracting resources and making things for household use, whereas workshop production is manufacturing for use elsewhere (Clark 1986; Hirth 2009a, 2009b). Household workshops producing surplus goods are not necessarily geographically coterminous with the house location but could be at resource locations. Distinctions often are made between independent production for the subsistence economy and attached production for the elite as part of the political economy. However, surplus household craft production evidently contributed significantly to the flow of goods and resources to elites and commoners by marketplace trade and exotics, and highly crafted goods made their way to commoners through markets as well (Hirth 2009a, 2009b; Masson et al. 2016).

Production involves acquiring raw materials and creating goods whereas specialization refers to how production is organized. Following Cathy Costin (1991), the organization of production often is evaluated by factors of intensity (part time / full time), context (independent/attached), concentration (dispersed to nucleated), constitution (kin-based to industrial), and scale (size). Since large quantities of materials can be produced by part-time, surplus household production, relative densities of production debris can be a useful measure of the intensity of production (Masson et al. 2016).

Production and specialization also are evaluated in terms of the number of crafts practiced and the periodicity of crafting (Hirth 2009b). Categories include intermittent single craft production, intermittent crafting, continuous single craft production, and continuous multicrafting. Intermittent crafting is conducted within the household along with subsistence activities and is often of short duration or periodic. A multicrafting household engages in the production of multiple crafts at the same time. Within multicrafting is a form called contingent crafting that represents the manufacture of products used in the larger crafting process. Salt production at the Paynes Creek Salt Works can be discussed as contingent multicrafting since pottery making, building construction, and other crafts were essential to salt production.

Elite craft workers at Maya royal courts included skilled specialists in writing, painting, and carving, often using imported obsidian, jadeite, or other materials that added value as "prestige" items. These goods were largely for royal gifting, exchange, and display that supported, maintained, and helped create the dynastic leaders' authority. Royal court artisans included highly skilled craft workers who produced painted pots, jade ornaments, painted murals, and sculpture, as well as scribes who painted hieroglyphic texts in paper books. Goods were produced for the dynastic leaders, who used them as gifts to their peers in other Maya polities and to lesser lords within their own polity, in royal feasts, and in other affairs of state. Sometimes such objects are found in burials; their use is recorded on pictorial painted pots and carved or painted on walls, such as the bundles of cacao on murals at Bonampak. The use of exotic materials added value to highly crafted goods. An impressive array of resources was obtained from distant locations, including jadeite, obsidian, basalt, marine shells, stingray spines, and mercury.

Food, materials for tools, and other subsistence goods and resources generally were obtained closer to home. However, the elite evidently acquired marine and other resources from farther away, albeit in smaller quantities than local foods (McKillop 1984, 1985). Fish must have been salted to maintain freshness for transport. Elite craft workers traded their produced goods in addition to commissioned goods for royalty (Inomata 2007). Excavations of structures in the central portion of Aguateca indicate attached elite craft production on a part-time basis for luxury items including bone and shell carving and wood carving. Attached workshops of elite households at Tikal (AD 250–850) provide evidence for the production of stone, shell, and bone goods (Moholy-Nagy 1997).

Ancient Maya workshops, including elite craft workers attached to the royal courts and independent household workshops, typically have production debris but few finished products (Andrieu 2013; Gonlin 2004; Guernsey 2016; Inomata 2001; Jordan and Prufer 2017; Martindale Johnson 2014; Masson 2001; Masson and Freidel 2012; Moholy-Nagy 1997; Robin 2012; Woodfill et al. 2015). Kazuo Aoyama (2017) describes lithic debris from various stages of producing obsidian blades at Ceibal. Workshops at

Tikal are identified by redeposited waste, with the workshops kept clean (Moholy-Nagy 1997). Workshop surfaces are sometimes identified by microdebitage in floors (Moholy-Nagy 1990) or crushed obsidian and shell in floors, as at Copan (Fash 2001). Workshops at Mayapán tended to have more standardized production intended for trade in contrast to household production for use as well as different types of chert tools and higher densities of production debris (Masson et al. 2016; Masson and Freidel 2012). Classic Maya workshops were located in houses or yards as well as near sources of raw material, such as chert outcrops at Colha, Xunantunich, El Pilar, and Chan (Hearth 2012; Horowitz 2017; Shafer and Hester 1983; VandenBosch et al. 2010; Whittaker et al. 2009), granite and slate near Pacbitun (Healy et al. 1995), and inland salt springs or the coast for salty water (Andrews 1983; McKillop 2005a; Woodfill et al. 2015). Commoners also were involved in the extraction and production of valuable materials such as jadeite (Rochette 2009), traditionally thought to be an elite good. By the Late Postclassic at Mayapán "most surplus craft workers were affluent commoners" working in households throughout the city, in addition to elite craftspeople and artisans working under direction of elites (Masson et al. 2016:229).

Marketplace Trade

The notion that some Maya plazas were used as marketplaces has long been suggested but has gained acceptance through identification of market stalls associated with plazas, chemical analyses of residues in plaza markets, and distributional studies of trade goods in communities. The depictions of vendors and their wares, including corn, chocolate, salt, and other items painted on the exterior of a building in the market area of Calakmul, Mexico, is compelling evidence of the importance of marketplace trade (Carrasco et al. 2009). The Classic Maya had marketplaces where vendors and shoppers congregated periodically. The marketplace at Tikal has two lines of masonry stalls in the center of a large plaza as well as other stalls and related market buildings (Jones 2015). Low, long mounds near city centers at other Classic Maya cities are viewed as locations of possible marketplace stalls at Maax Na, in northern Belize (King 2015). Marketplaces at the termini causeways at Caracol provided residential neighborhoods within the city with trade goods (Chase et al. 2015). The causeways provided pathways for traders, connecting outlying residential areas to the city-center at Caracol. Chemical analyses of plaza soils at Caracol, Chunchucmil, and Cobá have identified areas with evidence of food and craft activities, with supporting evidence from chemical analyses of soils at a modern marketplace (Coronel et al. 2015; Dahlin et al. 2007; Hutson et al. 2017; Terry et al. 2015). Evidence for a marketplace at Buenavista, Belize, was developed from subsurface remote sensing, soil testing, and analyses of micro- and macro-artifacts from shovel tests (Cap 2015).

The widespread distribution of exotic and other nonlocal goods to commoners, instead of their restriction to the dynastic leaders and other elite associated with the royal compound, underscores that market forces were at play (Masson and Freidel 2012). Carol Smith's (1974) model of periodic markets regularly scheduled at different villages in the Maya Highlands during historic times has been proposed for the ancient Maya (Freidel 1981) but has gained renewed interest for investigating the Classic Maya from the perspective of market economies (Eppich and Freidel 2015). Widespread distribution of fine ware pottery in the Late and Terminal Classic periods at El Perú-Waka' indicates equal access by the Maya at this city to goods via a marketplace (Eppich and Freidel 2015). Similarly, exotics as well as nicely made pottery traded from the Belize Valley (Belize Red) were distributed throughout Caracol to commoner households through marketplace trade (Chase and Chase 2012; Chase et al. 2015).

The degree of dynastic control of production and distribution of goods and services in the Maya area certainly varied over time and geographic space. In some cases, the elite may have taxed vendors or otherwise administered marketplaces (Braswell 2010; Chase et al. 2015; Masson et al. 2016:230). Taxation of surplus household production at marketplaces would have been a familiar concept to the dynastic Maya, who regularly exacted tribute, but marketplace fees may have been used instead to favor buyers and access to trade goods. Extensive part-time craft production by households contributed significantly to the amount of goods distributed throughout all levels of society, which also stabilized the political economy (Chase et al. 2015; Masson et al. 2016:230), was good risk management for households (Hirth 2009b), and may have been useful for distributing food in times of local shortages or surpluses (Eppich and Freidel 2015). Providing access to trade goods at marketplaces was a good management strategy by the dynastic Maya at Caracol. That commoners had choices of markets from which to obtain goods is also underscored by Marilyn Masson and David Freidel (2012:478), who envisage "nested market networks" with markets at major and secondary cities. Keith Eppich and Freidel (2015) suggest that food was traded at markets, with access to several markets both inside and beyond a city-state's region. Marketplace trade of food was a good risk-management strategy for times of food shortage in one area and surplus in another (Baron 2018a:110). Ethnographic data on travel distance to market of 18–30 km would take the Maya outside regions, an idea supported by the ceramic spheres that overlap regional boundaries (Eppich and Freidel 2015:Figure 5.1). Canoe travel would have extended that travel distance.

Currency Equivalencies

Goods were purchased with currency equivalencies such as cacao beans, lengths of woven cloth, copper bells, or obsidian blades (Baron 2018a, 2018b; Freidel et al. 2016; Tozzer 1941). The market economy was based on supply and demand that extended beyond the physical marketplaces and formed the structure of the Maya economy. The introduction of formed spindle whorls in the Late Classic is attributed to a need to standardize the quality of cotton cloth for exchange in the market, in contrast to perforated potsherd disks that continued to be used for household cloth production (Baron 2018a:108). Freidel and Justine Shaw (2000) suggest the use of currencies may have been a way to purchase food during shortages.

Ancient and Modern Salt Works

Paleoindian and Archaic people in the Maya area (Lohse et al. 2006; Prufer et al. 2016; Rosenswig 2015; Stemp et al. 2016) obtained sufficient dietary salt from animal meat. Salt production began toward the end of the Archaic period with the increase in horticultural practices and a decrease in the reliance on hunting. Evidence of salt production at most early sites is lacking, but salt production has been identified from the Late Preclassic (300 BC–AD 300) at Cerros in northern Belize (Robertson 2016). Oxtankah, located across Chetumal Bay from Cerros, was a large settlement that controlled Late Preclassic salt trade from 11 nearby salt works, using the site on South Tamalcab Island as a port (Reese-Taylor 2016). There is Middle Preclassic settlement at Salinas de los Nueve Cerros (Woodfill et al. 2015), so salt making may also date to that early time. More salt-making sites in the Maya area are known from the Classic period (AD 300–900), particularly to the Late and Terminal Classic (AD 600–900), when the demand for salt increased with population growth. The salt works expanded or arose in response to increased demand for salt to meet dietary requirements of larger populations inland (McKillop 2005a).

The organization of salt production varied tremendously in antiquity. The carbohydrate staples that underwrote dense urban populations in ancient civilizations were deficient in dietary salt, prompting quests for salt from mines, solar evaporation, and boiling brine from salt springs and from salt-enriched seawater. Some ancient civilizations controlled the production and distribution of salt by assigning administrators at salt works or by levying a salt tax, as in ancient China (Adshead 1992). State interest in salt can be documented by overseers' residences at salt works (Flad 2011; Li and Falkenhausen 2006). State interest in salt sometimes went beyond the dietary needs of the citizens, to providing state finances through taxation or tribute (Vogel 2010). During the Roman Empire, the state controlled the trade of, but not the production of, salt, with estimates of about 350,000 tons per year needed to provide dietary salt for about 50 million people (Saile 2015:202–204). Salt was part of the "salary" paid to soldiers. The Inca kept salt in state warehouses located throughout the empire and kept records on quipus of the amount of salt (DeLeonardis 2011). In the Sahara, salt was worth its weight in gold to the salt traders who drove camel caravans across the desert from inland salt mines (Benanav 2006).

Aztec salt production along Lake Texcoco is described as year-round, large scale, and specialized (Parsons 2001:300) or a "highly skilled craft activity that was practiced intermittently . . . alongside other subsistence activities . . . in the dry season and away from residential structures" (De Leon 2009:45). Even large salt works on the southeast coast of China in Shandong Province were evidently not large-scale, intensive producers of salt because only a few salt-making localities were used at a time (Flad 2013). Still, production was focused on salt, with over 90 percent of the artifacts salt-making briquetage. Residences were located inland from the coast, and overseers' residences were farther back (Flad 2013). At Zhongba, along the Yangtze River in China, the emphasis was on massive production of salt, but that focus later changed to producing salt to preserve fish (Flad 2007, 2011). Salt making in the Maya area along the coast of Belize often was associated with household production (Aimers et al. 2016; Valdez and Mock 1991). The exception is found at Late Preclassic Cerros, where there was a salt production and storage facility with salt produced for regional distribution and as an exotic alternative to local salt farther north (Robertson 2016:138, 143–144).

A Case Study of the Paynes Creek Salt Works

In this book I evaluate the scale and organization of Classic Maya salt production at the Paynes Creek Salt Works and its impact on supplying the nearby inland Maya salt needs (Figure 1.2). The Paynes Creek Salt Works are located at least 24 km from the nearest urban center; this distance suggests that the dynastic Maya rulers at inland urban centers did not directly control the production and distribution of this basic dietary resource, although they may have indirectly exerted control by placing a local representative at the salt work, as in ancient China, or by taxing salt at the salt works, or by overseeing inland marketplaces (Braswell 2010; Chase et al. 2015; Flad 2011). Salt production in the Maya area has been described as household production at Postclassic Wild Cane Cay and Frenchman's Cay (McKillop 2002) to possible elite-managed production at Salinas de los Nueve Cerros (based on public architecture beside the salt works; Woodfill et al. 2015). Trade goods—notably, pottery—at the Paynes Creek Salt Works may indicate the communities to which the salt producers traveled to sell salt, salted fish, and other marine resources at markets. The salt producer–vendors may have returned home with nicely made pottery and other goods and resources, including corn.

The Paynes Creek Salt Works consist of 110 known sites, mainly in a 5 sq km area, with a few isolated salt works beyond. If they were all used at the same time, this would have been a significant level of production. In contrast, if only one or a few salt works were in use at a time over several hundred years, this would have been small-scale production with less impact on the supply of salt for inland use. Preservation of ancient Maya wooden buildings provides a window into the nature of the salt industry elsewhere along the coast of Belize where wood has not preserved. The interpretation of the process of salt production before the discovery of the wooden buildings was that production was outdoors (Mosnier 1994). Discovery of the wooden buildings indicates more intense production in

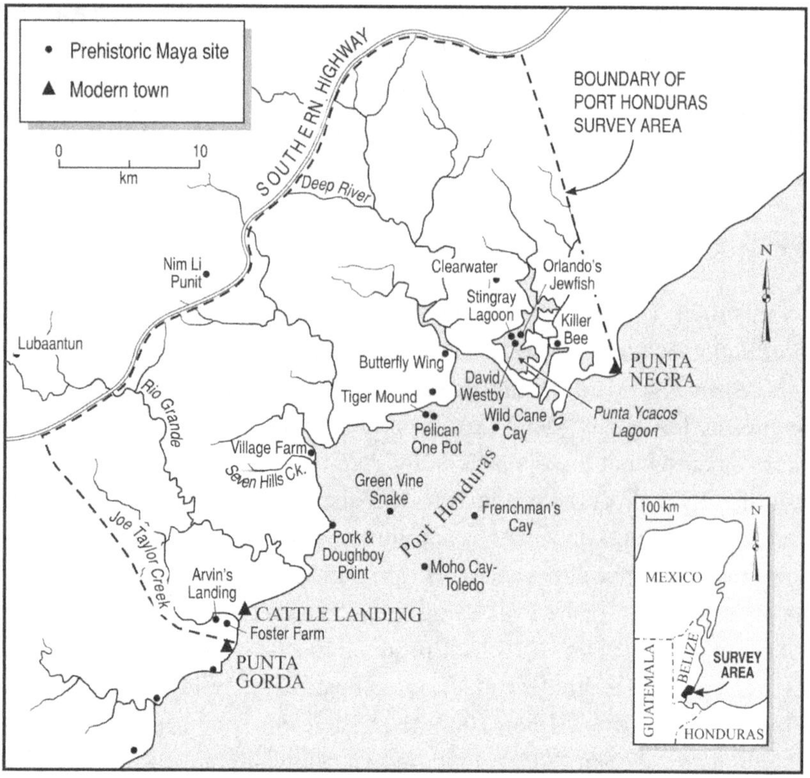

Figure 1.2. Port Honduras and Punta Ycacos Lagoon survey area, with inset showing location of map within the Maya area. By Mary Lee Eggart.

salt kitchens, similar to modern salt production at Sacapulas in the Guatemalan highlands (Reina and Monaghan 1981).

The building posts were preserved where they had been driven during construction into the ground, which was anaerobic mangrove peat produced as red mangroves kept pace with rising sea levels (McKillop et al. 2010a, 2010b). Global sea-level rise since the end of the Pleistocene inundated low-lying coastal areas. In Belize, the barrier reef system developed during the last 10,000 years, resulting in the modern coastal landscape with a shallow inshore lagoon containing hundreds of cays. The barrier reef extends up to 40 km offshore in southern Belize. As much as 11 m of mangrove peat has been documented below the sea floor within the barrier lagoon in southern Belize, reflecting sea-level rise. Land subsidence also contributed to relative sea-level rise.

The preservation of wooden architectural posts and beams, together with briquetage and charcoal directly on the ground without wooden floors, supports an interpretation that the buildings were constructed on land. The wooden posts were preserved because they were driven into the mangrove peat. Had the posts supported wooden structures above the water, the posts would show deterioration from burrowing sea worms, barnacles, and from exposure to the air. The salt works developed to satisfy the need for salt and salted fish by the inland Maya. When the nearby inland Classic Maya cities were abandoned, the Paynes Creek Salt Works were abandoned for lack of customers—although this may not have occurred immediately since people still lived on the coast and offshore islands. Sea-level rise eventually drowned the salt works—a sobering reminder for low-lying coastal areas worldwide.

Other ancient Maya communities that were suddenly abandoned provide a useful analogy for the Paynes Creek Salt Works, including Aguateca, Guatemala (Inomata et al. 2002), and Ceren, El Salvador (Sheets et al. 2015). The Classic Maya village of Ceren was preserved by tephra from a volcanic eruption that destroyed the community. Aguateca was invaded and burned, and although the elite were able to flee, they left virtually everything in their homes. Takeshi Inomata and his colleagues piece-plotted individual artifacts on burnt floors of elite residences. The team also mapped artifacts that may have been left by the Maya on shelves or in the rafters but fell to the floor when the roofs and walls collapsed. The artifacts did not indicate a moment frozen in time; rather, the objects likely were left in storage in the rooms in which they were used. A similar situation likely occurred at the Paynes Creek Salt Works, with sea-level rise protecting artifacts from postdepositional trampling by animals or later visitors.

The wooden structures, along with the first reported ancient Maya wooden canoe paddle from the K'ak' Naab' underwater site, were discovered in 2004 (McKillop 2005a). Survey from 2005 to 2009 resulted in the discovery and mapping of 4,042 wooden posts and other architectural wood defining rectangular buildings as well as other structures that may have been platforms where concentrated brine was produced by pouring water through salty soil and collecting the salty water below prior to the evaporation process in pots over fires (McKillop 2005a, 2010a, 2011, 2012, 2015, 2016a; McKillop and Sills 2017; Sills and McKillop 2010; Somers 2007). Since stone buildings and the mounded remains of perishable structures dominate the modern landscape of Maya sites, the Paynes Creek wooden structures provide a key to documenting ancient wooden buildings, to evaluating analogies with modern structures, and for providing analogues for other ancient sites.

The current study builds on previous research on four salt works in the same lagoon system where pottery vessels were standardized in their dimensions, indicating mass production of salt (McKillop 2002). Standardized pottery and buildings represent an organized effort to produce salt for marketplace trade. The Stingray Lagoon, the David Westby site, and the Orlando's site had briquetage that was standardized in dimensions in contrast to household pottery from the nearby island trading port and community at Wild Cane Cay (McKillop 2002:127–132, Tables 4.1 and 4.5). The salt works in that study also had briquetage with dimensions statistically different from one another, suggesting different work parties, perhaps different households. Was this pattern typical for the Paynes Creek Salt Works? For the larger sample of sites discussed in this book, in addition to the standardization of briquetage, I evaluate the sizes and tree species of wood posts within and among salt works to see if they were standardized and to see how the results compare with the briquetage.

The Paynes Creek Salt Works produced and supplied salt within the complex Classic Maya economy of tribute, marketplaces, gifting, and reciprocal exchange. Perhaps canoe loads of salt, other marine resources, and cacao from the rich coastal farmland were sent up rivers to markets at Nim Li Punit, Lubaantun, and elsewhere. They would have returned with goods and resources from the market or currency equivalencies to acquire other commodities later. Salt traders may have returned with gifts from dynastic leaders, including ritual items or valuables, to further integrate and maintain the social and economic bonds to ensure a regular supply of salt. Perhaps inland Maya ventured seasonally in the nonfarming dry season to the Paynes Creek Salt Works to make salt to take back to inland markets. Alternatively, coastal Maya from Wild Cane Cay, Village Farm, or other communities near the Paynes Creek Salt Works paddled daily to the household or community-owned salt works in the lagoon system. Certainly, the recovery of canoe paddles, a canoe, and various small boat models underscores the use of canoe travel. The ease of water transport, especially for bulk items such as salt cakes, salted fish, and bags of cacao seeds, further suggests water transport was used. The coastal

Maya were mariners who were familiar with the shoals, storms, and currents that pose dangers on the open sea (McKillop 2007a, 2010b, 2017b), skills not known to the inland Maya, further underscoring the likelihood that commodities from the Paynes Creek Salt Works were transported by the coastal Maya up rivers to inland markets.

This book includes evaluation of the economics of salt production and trade from analyses of the spatial patterning of wooden buildings and associated artifacts that were mapped on the sea floor at the Paynes Creek Salt Works. Chapter 2 describes methods developed from terrestrial and underwater archaeology for use in shallow, underwater sites. Chapter 3 discusses the technology of salt production using the brine-boiling method. This includes the pottery vessels used to boil the brine as well as the brine-enrichment and transport in pipes to the salt kitchens. The value of ethnographic analogy, ethnoarchaeology, and historical and cross-cultural comparisons is underscored for interpreting salt production methods. Chapter 4 describes the wooden posts, beams, and wedges and their spatial arrangements as salt kitchens. Tree-species selection for use in building construction is evaluated in chapter 5 in the five areas of the lagoon system in relation to the distinct forest zones nearby, including the mangrove ecosystem, coconut woodland, pine savanna, and deciduous rainforest.

The remaining chapters are synthetic, building on information provided in the first chapters. Chapter 6, "Labor, Production, and Identity," includes a discussion of the organization of salt production using information about the briquetage and salt kitchens. The data include briquetage densities and standardization of salt containers, tree-species selection for salt kitchens, and the use of nonbriquetage ocarinas (figurine whistles) to identify salt-worker households. Chapter 7 identifies the inland consumers of dietary salt using nonbriquetage trade pottery as a proxy. The discussion is set within a framework of pan-lowland marketplace trade built on surplus household production. Going beyond the biological importance of dietary salt, standardized units of salt cakes are proposed as a valuable commodity in market exchange, as a storable commodity with further value as a currency equivalency, and as stored wealth. Trade in salted fish, a storable food, is discussed from the evidence that significant amounts of salted fish were prepared at the Paynes Creek Salt Works. Chapter 8, "Salt: A Mover and Shaker in Ancient Maya Society," estimates the quantity of salt produced at the Paynes Creek Salt Works and the dietary salt needs at nearby inland communities in southern Belize using ethnographic data. A model of the role of salt in the Maya economy is proposed, consisting of surplus household salt production using various methods that was integrated in marketplace trade by the Late Classic period.

2

Discovery and Mapping Underwater

Low-lying mangrove islands and coastal lagoons characteristic of mangrove ecosystems dominate the modern coastal landscape of southern Belize (Figure 2.1). Observations of artifacts in the shallow water around Wild Cane Cay led to stratigraphic excavations in the offshore area, resulting in the discovery of deeply buried, intact archaeological deposits below the sea floor (see Figure 2.2; McKillop 2002:150–156, Figures 5.2–5.6; 2005b). Regional surveys in Port Honduras revealed that Classic Maya sites are hidden in the modern landscape, which has been transformed by sea-level rise, changing the vegetation patterns and reducing the amount of land (McKillop 2002:135–174). Sea-level rise submerged Classic Maya settlements on the southern coast and offshore cays (islands) of Belize. Most of the ancient sites in the area are either underwater or buried below the ground surface and partly or completely below the modern water table (Figure 1.2; McKillop 1996a, 2002). Excavations on mangrove islands also revealed archaeological sites buried below mangrove peat, such as at Pelican Cay (McKillop 2002:156–159, Figures 5.5–5.6).

There are four sites with stone architecture, four sites with earthen mounds, and one shell midden site known in the coastal area. In addition, there are six underwater sites with shell middens at the Paynes Creek Salt Works (Feathers et al. 2017; McKillop 2017a) and shell middens at Wild Cane Cay and Frenchman's Cay (McKillop 2005b; McKillop and Winemiller 2004). Coral rock was used as foundations for pole-and-thatch buildings at Wild Cane Cay, Frenchman's Cay, and Green Vine Snake, all islands in Port Honduras adjacent to Paynes Creek National Park (Figure 1.2). Chert river cobbles were used to make a building platform at Arvin's Landing, on Joe Taylor Creek in Punta Gorda. Tiger Mound has an earthen mound in a clearing characterized by dead mangroves and salty soil in the middle of the island. There are deeply buried Classic period deposits in

Figure 2.1. Oblique air view of the Main Channel and East Lagoon area of the Paynes Creek Salt Works.

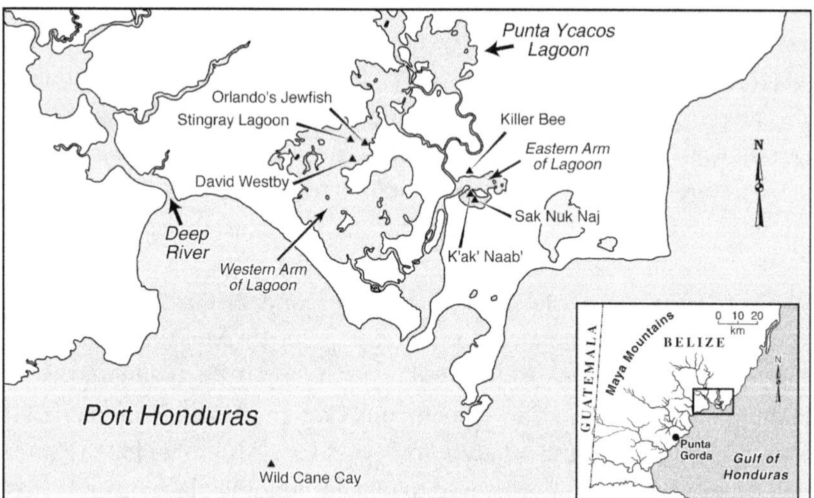

Figure 2.2. Punta Ycacos Lagoon showing location of selected sites. By Mary Lee Eggart.

the mangroves. Seven Hills has an earthen mound. Two of the Paynes Creek Salt Works have earthen mounds that were slag heaps from leaching salty water through salty soil to enrich the brine before the evaporation process (McKillop 2002; Watson 2015; Watson et al. 2013). Perhaps the earthen mounds at Tiger Mound and Seven Hills also were part of the salt-making process. Ich'ak'tun is a shell-midden site dated from the early Middle Preclassic to Late Preclassic / Early Classic transition (McKillop and Robertson 2018). Other sites with shell middens include Wild Cane Cay, Frenchman's Cay, and several sites at the Paynes Creek Salt Works (Feathers and McKillop 2018; Feathers et al. 2017; McKillop 2005b, 2017a; McKillop and Winemiller 2004).

With the knowledge that Late Classic Maya sites were buried at least one meter below the modern water table, as demonstrated by radiocarbon-dated submerged deposits (McKillop 2002:Table 5.3), survey was initiated in shallow water, including Punta Ycacos Lagoon, the large salt water lagoon in Paynes Creek National Park. Three underwater sites were discovered in 1991 in Punta Ycacos Lagoon, a large salt water lagoon system in Paynes Creek National Park, including Stingray Lagoon, David Westby, and Orlando's (Figure 2.2). The sites were excavated in 1991 and 1994 (Braud 1996; McKillop 1995, 1996a, 2002). A fourth site, Killer Bee, was visited in 1988 with former local resident Charlie Carsen in his boat, the *Native Gal*. The sites were marked by the presence of ceramic artifacts on the sea floor or, in the case of the Killer Bee site, by the presence of artifacts in the black mangroves and on a low, earthen mound.

The sites were identified as salt works by ethnographic analogy as workshops where brine was evaporated in pots over fires to produce salt, as at Sacapulas, Guatemala (Reina and Monaghan 1981) and elsewhere, including Thailand and the Philippines (Halliwell et al. 2016; Yankowski 2008, 2010; Yankowski and Kerdsap 2013; Yankowski et al. 2015). Salt was hardened in standard-sized jars to produce salt cakes at San Mateo Ixtatán, in the highlands of Guatemala (Andrews 1983:Figures 4.9–4.10) for market trade. On the island of Bohol in the Philippines, hardened salt cakes were produced and traded in standard-sized jars as a unit of exchange (Yankowski 2010). Different villages used different sized pots for their trade. Sometimes salt cakes in pots were split in half for sale.

Although the salt-making pots at the Paynes Creek Salt Works were standardized in size suggesting surplus household production of salt for trade (McKillop 2002:127–132; Tables 4.1–4.5), the nagging question was whether there were more salt works that might have satisfied the dietary salt needs of the nearby inland Maya. Survey was planned to discover if there were more salt works in order to evaluate the potential contribution of salt trade between the coast of Belize and inland cities of the Southern Maya Lowlands during the Late Classic period. Five days of pedestrian survey in the water in 2003 resulted in the discovery of six additional sites, underscoring the viability of an underwater search.

Systematic survey was carried out in the lagoon with a team of three to four people using the terrestrial archaeological technique of pedestrian survey: The team walked at arm's width apart, crisscrossing sections of the lagoon to examine the entire sea floor. Water depth was 1 m at most, with deeper areas avoided. Sites were identified by the presence of pottery. The location of underwater sites was recorded using a GPS.

Underwater sites with wooden architecture were discovered in 2004 during a pilot project to systematically survey Punta Ycacos Lagoon. The first building post was discovered at Site 15 after excavating a piece of wood that barely protruded above the sea floor. Excavation around the wood into the peat indicated that the post was buried about a meter below the sea floor and had a sharpened end at the base (Figure 2.3). Searching near the post revealed additional posts that outlined a rectangular structure, associated with briquetage, so the building was referred to as a "salt shed," and later a "salt kitchen," following terminology used by modern Maya salt makers in the highlands of Guatemala (Reina and Monaghan 1981). Returning to the 14 sites identified as such by the presence of briquetage, wooden posts were discovered at all sites. Posts also were found at the underwater sites initially discovered and excavated (McKillop 2011). A transit was set up in the lagoon at a datum marked by a 4-foot length of ½-inch PVC pipe in order to begin mapping the distribution of remains at discovered sites.

Wooden architecture was preserved in locations where the Classic Maya sunk building posts into mangrove peat that was exposed above the water due to fluctuations in sea level. In northern Belize, limestone bedrock is near the ground surface. In contrast, the deeper waters of southern Belize allowed extensive mangrove peat formation during the Holocene (Macintyre et al. 1995; McKillop 2002). Mangrove peat extends below the sea floor between the coast and barrier reef in much of southern Belize. Fluctuations in sea level during antiquity resulted in some areas of the sea floor in southern Belize being exposed above the water.

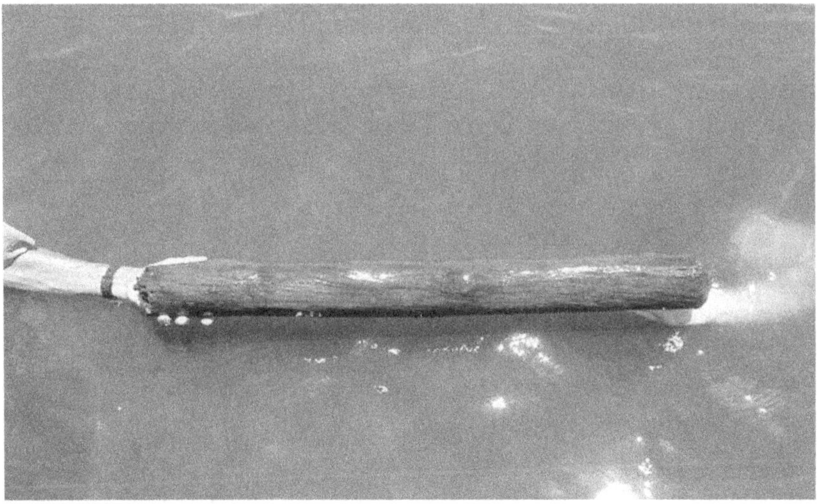

Figure 2.3. Wooden post found at Site 15 in 2004.

The Classic Maya constructed salt kitchens along the shoreline of the salt water lagoon in what is now Paynes Creek National Park. During the dry season, the water was saltier than the open sea due to solar evaporation, which reduced fuel needed to evaporate brine in pots over fires to produce salt. When the Classic Maya constructed salt kitchens on this land, they drove the building posts into the ground, which was composed of mangrove peat. The peat preserved the wooden posts and other wooden objects and plant remains, such as palm fronds, leaves, and seeds. The coincidental location of the sites in a peat bog below the sea floor preserved the wood. Sea-level rise resulted in submergence of the salt works, leaving most hidden from modern view below the water.

Potsherds and vessel supports from bowls and jars used in the salt-making process were abundant at the 33 underwater sites discovered in 2004. Together with four sites previously reported (McKillop 1995) and eight sites discovered in 2003, the total of 45 salt works in Paynes Creek indicates that a significant amount of salt was produced in the Late Classic on the southern coast of Belize. This finding, along with the presence of unit-stamped pottery and clay figurine whistles from inland sites at the salt works, reinforces the significance of coastal–inland trade of salt and other commodities in southern Belize at the height of the Classic Maya civilization (McKillop 2005a).

The Search for Wooden Architecture below the Sea Floor

Field research was carried out between 2005 and 2008 to map the distribution of wooden posts and other wooden structural remains visible on the sea floor at the underwater sites in Paynes Creek National Park, to continue the systematic underwater search, and to analyze the spatial distribution of wooden structures and other remains. The research team included the boat driver who also participated in the fieldwork, two to four graduate and undergraduate students, and me. The maximum team size was restricted by the boat capacity.

The only permanent residents of the area invited the team to stay with them at their home located at Village Farm, midway between Punta Gorda and the Paynes Creek Salt Works. Paynes Creek National Park is uninhabited except for a ranger station. The nearest community, Punta Gorda, is located some 30 miles by water. There are no roads. The only way to access the research location is by boat, either private boat or boat charters. The team commuted daily from Village Farm to the salt works, with lab research at Village Farm. Access to electricity from a gas-powered generator was used to recharge field equipment, operate electronic equipment, and download data and images. We were able to enter map data into the project geographic information system (GIS), Underwater Maya, create and update maps, evaluate patterning of posts, and use the distributional data to identify locations to search for missed posts (McKillop 2012).

Fieldwork was carried out during the dry seasons, with additional field and lab work during the rainy season, June to August. The optimal time of year for fieldwork in the lagoon is April to the end of May. During the rainy season (mid-June through February), visibility in the water is negligible due to turbidity of the water from silt brought down rivers. Moreover, the seas are rough, and rain and thunderstorms are hazardous for fieldwork in the water. In 2008 the early onset of the rainy season at the end of May was marked by the unexpected arrival of Tropical Storm (TS) Arthur from the Pacific, followed by the unexpected arrival of TS Alma in the Atlantic. We experienced heavy rain and bands of strong winds, with zero visibility during TS Arthur. We were unaware of TS Alma, which developed suddenly during the night. Commercial radio communication was cut. Our phone and ham radios were turned off due to thunderstorms. Parts of Belize were extensively damaged from flooding, including the Sittee River steel bridge washing out, preventing road

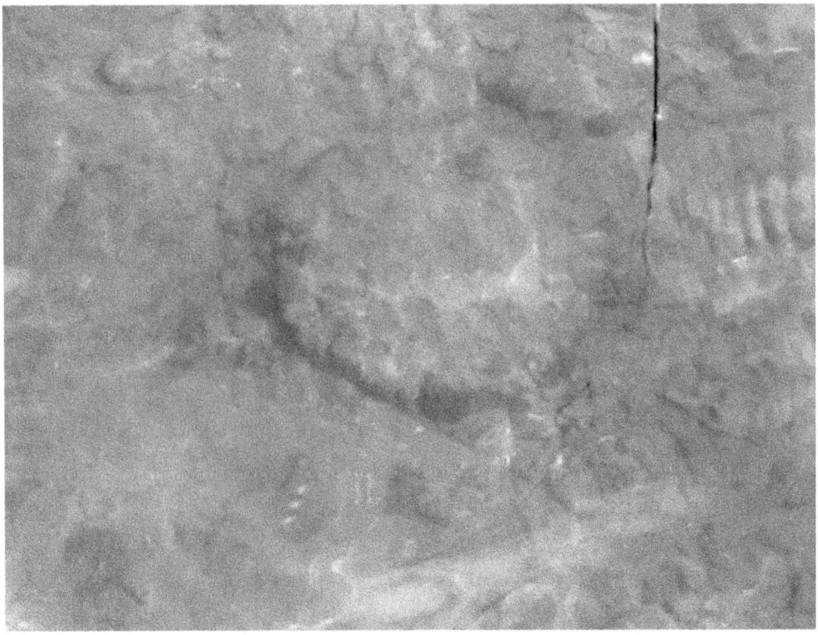

Figure 2.4. Underwater image of the top of a wooden post protruding from the sea floor, with a wire flag marking the post location.

travel to and from the Toledo District. Hurricane Iris cut an estimated 40 percent of mature rainforest at Village Farm on October 8, 2001.

Although shallow underwater sites are regarded by underwater archaeologists as difficult to map and excavate (Purdy 2001), we figured out a system to successfully survey and map the Paynes Creek underwater wooden buildings: The land-based archaeological technique of pedestrian survey was used in a modified form, with the team snorkeling on research flotation devices (RFDs), shoulder to shoulder, systematically snorkeling back and forth across each site, looking for wooden posts and other artifacts on the sea floor (McKillop 2016a). Many artifacts were buried under deep silt and embedded in the sea floor, so they were invisible during flotation survey and could only be discovered by feeling the sea floor by hand. Wooden posts were difficult to see since they barely protruded, if at all, above the sea floor (Figure 2.4). Artifacts and posts were marked by wire flags, with a flag placed on the north side of finds, so they could be easily relocated underwater. Palmetto palm posts, in particular, were rarely visible. They normally were discovered by feeling the thin, exterior shell of the post, since the interior wood decayed near the surface of the sea floor, only preserved deeper. In fact, we had not seen hundreds of wooden posts at Stingray Lagoon (McKillop 1995), David Westby, and Orlando's (McKillop 2002) during survey and excavations. Not only are the posts difficult to find, we did not expect to find them because wood does not normally preserve in the tropical landscape of the Maya area (McKillop 2011).

The flotation survey was modified so that each surveyor placed their hands flat on the sea floor, moving them to their left and right to feel for artifacts, wooden posts, and the shell exteriors of palmetto palm posts. Hands, fingers, and fingernails were used to dig into the sea floor to define the edges of artifacts and posts. Once the systematic flotation survey of a site was completed, the team members searched for additional posts that were not discovered, often buried below the sea floor. Palmetto palm posts often formed long lines, so searches followed posts along a line. Other wooden posts often formed the footprints of rectangular buildings, so gaps in this pattern were searched. This search was carried out by looking at the flags that marked the locations of posts and from paper maps of the post locations produced by the GIS, when available. Sites were mapped with a total station. In the evening at our base at Village Farm, using power from a generator, the digital data were downloaded to a laptop, transferred to an Excel file, and attached to the project GIS, Underwater Maya. Paper site maps were printed and placed in heavy-duty ziplock bags, attached to a clipboard and used as a reference for further survey. The combination of systematic survey and filling in the gaps was successful in post discovery, especially since many posts were buried below the sea floor. Few posts were visible even by touch.

Floating above the sea floor improves visibility of the sea floor since the bottom silt is not disturbed and additionally preserves the integrity of the sites (since walking leaves footprints, or worse still, gaping holes into the sea floor). Initially the team wore masks and snorkels, but some team members became accustomed to the salt water and opening their eyes underwater. All team members wore polarized sunglasses out of the water for protection from ultraviolet rays, which were intensified from reflection from the water.

The survey team stayed in formation, stopping to mark posts and artifacts, and rotating in unison at the end of each traverse to begin the next traverse. Each end of a traverse was marked by a ½-inch PVC pipe long enough to protrude above the water. The PVC survey pipes were moved as the team traversed the site, leaving the survey area boundary to be mapped as surveyed. Wooden posts were marked for later mapping. Selected diagnostic or culturally significant artifacts were individually marked for mapping. Wire and plastic survey flags (24" for plastic and 36" for wire flags) were used in shallow water. Although useful since they are longer, wire flags rusted, making them hazardous. After mapping and cutting the tops of posts for species identification, flags were replaced with plastic flags furled inside plastic straws that we sunk into the sea floor beside each post. The post number was written in indelible marker on the flag before it was furled and three times on the side of each straw. Subsequent removal of selected straws revealed the marker writing remained if the straws had remained intact in the mangrove peat below the sea floor. This system was useful for relocating posts. In deeper water, we used fishing line, with florescent Styrofoam fish floats tied to one end and metal wires at the other end to drive into the sea floor. The fishing line was adjusted in length according to the tides so that they accurately marked the location of the post or artifact under water. Flags were placed on the north side of posts and artifacts for ease of relocation during mapping. North was verified when needed by a compass watch. To facilitate mapping and see patterns of underwater posts from the water surface, color-coded flags and flagging tape were used to distinguish posts and artifacts. Red flags were used to demarcate wooden posts. Yellow flags were used for artifacts. When fishing line was used, color-coded flagging tape was tied to the fish float.

Only some of the wooden posts are easily seen protruding from the sea floor. Since many of the wooden posts barely extend above the sea floor or are buried, and since they have the same texture as the peat, posts can easily be missed. Locating posts used a combined approach. The team looked ahead in the clear water for any posts protruding from the sea floor, which at most sites was covered by a layer of silt measuring from a few centimeters to 20 cm or more. The silt layer is deeper in the East Lagoon than parts of the West Lagoon where Hurricane Iris scoured the sea floor and moved the silt into the mangroves. There is deep silt in the western arm of the West Lagoon by Stingray Lagoon Site. Most posts were discovered by feeling the sea floor or digging with fingers slightly into the peat. Techniques varied from site to site depending on the amount of mangrove peat growth. Many sites were partly covered and obscured by modern growth of red mangroves. Sites 14, 15, 18, 19, and 72 had clumps of modern mangroves covering part of the sites. Site 16 is located in a "V" cove by the shore of red mangroves, and the site extends into the mangroves. Red mangroves bisect site 60, Ek Way Nal.

Mapping Underwater

Mapping improved over the course of the project, resulting from using better equipment and from fine-tuning our techniques. In 2004 we used a transit set up in the lagoon to map with a 100 m steel tape and stadia rod, which lacked precision and the data were discarded. In 2005 we used a total station to record distance, angle, and elevation in a notebook. The mapping team included one person at the instrument taking measurements and one person holding the prism pole at each post for artifacts. The team used walkie-talkies to communicate.

From 2006 to 2008 we used a total station in the field and in the evening at our field station downloaded the digital coordinate data and added the data to the project GIS. The 2005 data were converted to coordinate data and added to the GIS. Producing maps in the field allowed us to learn from the previous day's mapping to direct fieldwork: We returned to sites with paper copies of post distributions in ziplock bags to search for additional posts or to clarify the post patterns. For example, at Site 58 (Orlando's), we identified a triangular pattern of three posts, returning to the site to search for a fourth corner post, which was located, making a square pattern. Also at Site 58, we identified two areas of U-shaped lines of posts facing each other on the GIS map. Were these separate structures or had we missed intervening posts? We resurveyed between the two areas and found additional posts that indicated there was one large rectangular structure.

Sampling Posts, Plants, and Artifacts

Wood samples were cut from posts at different sites in order to date the wooden structures and to see what kind of wood was used in building construction. I applied for temporary export permits from the Belize government Institute of Archaeology for radiocarbon and species analyses of wood samples. Radiocarbon samples were sent to Beta Analytic. Modern wood samples were collected and exported for species identification to add to the author's existing modern reference wood collection at Louisiana State University.

Wooden artifacts and selected wooden posts were mapped and exported for conservation using the polymer process (Smith 2003) under permit from the Belize Institute of Archaeology. We left most of the posts in the peat bog, which is the best condition for their preservation. Posts were marked by plastic straws, whose location can be relocated by drawing one's hand above the surface of the sea floor. The straws replaced fluorescent orange flags on metal stems used during the fieldwork but were removed to deter looting at the sites.

The mangrove peat preserved plant remains, which were collected as indicators of past environments and food. The endocarps of native palm fruits from cohune (*Orbignya cohune*) and coyol (*Acrocomia mexicana*) were common on the surface of the sea floor. When found at a site, a sample was collected as an indicator of the past environment. Other plant remains were individually flagged, mapped, and collected.

Ceramics of interest were flagged and individually mapped. Ceramics regarded as temporally sensitive for building a type-variety classification were collected to help date the sites. In addition, examples of briquetage were flagged, mapped, and collected to broaden the inventory of vessel forms, supports, and other sand-tempered Punta Ycacos Unslipped ceramics. The ceramics were either dried and stored for study or kept wet if surface decoration required conservation. We carried out type-variety and attribute analysis at our field lab at Village Farm. Stone tools made from chert, obsidian, and ground stone were flagged, individually mapped, and collected. Relatively few stone artifacts were found in comparison to the briquetage. Selected artifacts were 3D imaged at the field camp, with files printed and used in exhibits featuring 3D-printed replicas of artifacts and posts in Punta Gorda and at the Paynes Creek Ranger Station (McKillop 2016a; McKillop and Sills 2013a, 2013b).

3

Briquetage and the Technology of Salt Production

The ancient Maya used two main methods of evaporation to produce salt, including solar evaporation and evaporation by heating in pots over fires. Solar evaporation was practiced on the north and west coasts of the Yucatán, at La Concordia where there is a salty lake in the highlands of Chiapas, and historically at some locations on the Pacific coast of Guatemala (Andrews 1983:Figure 3.2, 68–77). Inland salt springs were used to collect salty water that was boiled in salt kitchens at Sacapulas, San Mateo Ixtatán, Ixtápa, and other locations in the highlands; at Salinas de los Nueve Cerros; and at salt works along the coast of Belize and Pacific coast of Guatemala and Chiapas, Mexico (Andrews 1983; Coe and Flannery 1967; Nance 1992). This technique was used at the Paynes Creek Salt Works, based on ethnographic and ethnoarchaeological analogies and comparisons with other prehistoric salt works, especially Sacapulas, in the highlands of Guatemala (Reina and Monaghan 1981). Ethnoarchaeological and ethnographic studies of salt production are critical to developing models and middle-range theories for archaeological interpretation since, in both modern and ancient times, salt production included activities that leave few or no tangible traces. Brine boiling similar to that carried out at the Paynes Creek Salt Works is documented in case studies of preindustrial salt production worldwide, notably in Asia, Africa, Europe, and the Americas (Andrews 1983; Antonites 2016; Brigand and Weller 2015; Brown 1980; Dumas 2007; Eubanks 2013; Fu 2015; Gouletquer 1975; Halliwell et al. 2016; Li and Falkenhausen 2006, 2010; Masatoshi 2010; Olivier 2010; Parsons 2001; Reinecke 2010; Saile 2010; Sordoillet et al. 2018; Williams 1999, 2003; Yankowski 2008, 2010; Yankowski and Kerdsap 2013; Yankowski et al. 2015).

Salt was important in antiquity when the diet became reliant on salt-deficient, high-calorie carbohydrate grains including barley, corn, millet, rice, sorghum, and wheat. It is important to figure out how people obtained salt—geographically restricted in natural availability but universally required biologically and for many other uses. Salt became a commodity produced and packaged for wide distribution. Salt's value was enhanced as a preservative for fish and meat or as fish sauce—popular among the ancient Chinese, Thai, and Romans—as a good risk-management strategy for managing availability near and far (Yankowski et al. 2015:7–10). Fish sauce—the liquid derived from fermented fish—was regarded as an industry at Pompeii and was stored and transported in special containers (Curtis 1984). Fish sauce and other fish products may have been the fourth most important trade commodity, after wine, olive oil, and grain, in Roman times (Locker 2007). Packaged salt was a surplus commodity to be stored and used as needed—a currency equivalency—but also something in which the state took an interest to underwrite, either controlling production or distribution as in Roman and Chinese antiquity, especially as taxes, or even going to war over the control of salt works (Falkenhausen 2006; Saile 2015; Vogel 2010). The Roman state had a monopoly on wholesale trade of sea salt produced on the shores of the Mediterranean but left production in private hands (Saile 2015:Figure 1, 202). Sea-salt production by solar evaporation was expanded during the Roman Empire, with evidence of a Roman coastal salt work preserved at Vigo in Galicia.

Evidence of Maya salt production from Middle Preclassic origins through the rise of Classic period civilization in the Southern Lowlands and the Spanish contact and colonial periods reveals the methods and materials used to make the salt that underwrote the organization of production and distribution of this critical commodity (Andrews 1983; Kepecs 1998, 2005; Reese-Taylor 2016; Robertson 2016; Woodfill et al. 2015). The salt flats along the coast of the Yucatán are regarded as important sources of dietary salt as early as the Late Preclassic through modern times (Andrews 1983; Eaton 1978; Kepecs 1998, 2003a, 2003b, 2005; Sierra Sosa et al. 2014). Solar evaporation is possible because the area is arid with a long and predictable dry season (West and Augelli 1989). Rainfall increases moving south in the Yucatán, with high rainfall and short and unpredictable dry seasons in southern Belize, curtailing solar evaporation and often moving salt production indoors. Coastal salt flats extended across the entire northern coast within estuaries and along the western coast of the Yucatán and included major salt flats at Celestun and Emal. Coastal ports of Xcambó, Emal, and Vista Alegre may have facilitated distribution of salt at nearby salt flats (Kepecs 1998, 2003a, 2003b; Sierra Sosa et al. 2014). Terminal Classic expansion of the Chontalpa sea traders from the Laguna de Términos region of Campeche lead to their control of coastal salt works and establishment of their economic capital at Chichén Itzá (Andrews and Robles 1985; Kepecs 2003a). The control of coastal sea trade by Chichén Itzá was felt even farther away along the coast of Belize, where round structures, Yucatecan slate ware pottery, and Mexican obsidian are widely reported (Harrison-Buck and McAnany 2013; Masson and Mock 2004; McKillop 1996a, 2005b).

Salt pans are used for solar evaporation along the coasts of the Yucatán to cordon off sections of shallow estuaries but also on a smaller scale at inland lakes, such as La Concordia in the highlands of Chiapas, Mexico. Salt pans are also constructed on land, such as at La Placita in Michoacan, Mexico (Andrews 1983; Williams 2003). They are formed by lines of posts in the water or by baked earth or clay on land. They serve to increase the salinity of brine by solar evaporation. In the large salt works on the north coast of the Yucatán and Gulf of Nicoya in Costa Rica, lines of posts demarcate the edges of salt pans (Andrews 1983:Figure 2.3; Bishop 1977). Successive salt pans often are used to move brine during the evaporation process, with sluices between ponds, or by using pipes or containers to physically move brine.

Making salt by boiling brine in containers over a fire is common worldwide, historically and in ancient times. This technique often results in briquetage—the broken salt-making pottery—at archaeological sites. Briquetage is abundant and high density—over 95 percent of artifacts from some excavations—at the Paynes Creek Salt Works (McKillop and Sills 2016). Briquetage vessels were used to boil brine and, in some cases, as molds to make salt cakes to trade, as at Zhongba, China, and in the Maya Highlands (Flad 2011; Li and Falkenhausen 2006). Briquetage includes clay cylinders that were sometimes used to support the brine-boiling vessels over a fire as well as other related brine-boiling pottery and kilns. The term "briquetage" was coined in France in 1740 by Félix-Francois Le Royer Arteze de la Sauvagere, who observed brick-like mounds at Marsal and other sites in the Seille Valley of Lorraine, France, now known to date as early as the end of the Early Iron Age (500–50 BC; Falkenhausen 2006:23; Nenquin 1961; Olivier 2010). Dense deposits of briquetage were found at Bourthecourt, with an estimate of 4,000,000 cu m of briquetage up to 10 m in thickness in the Seille Valley, France (Olivier 2010; Olivier and Kovacik 2006).

The widespread occurrence of briquetage along the coast of Belize demonstrates that salt production involving brine evaporation in pots over fires was common in the Southern Maya Lowlands. Salinas de los Nueve Cerros dates as early as the Middle Preclassic (Woodfill et al. 2015). There is evidence of salt production in the Preclassic at Cerros (Robertson 2016) and across Chetumal Bay at 11 sites (Reese-Taylor 2016). There are other Late Preclassic salt works at Salinas de los Nueve Cerros, Ocós, Guzman, and near Ixtápa (Andrews 1983:56, 59–60). That most salt works with briquetage along the coast of Belize date to the Late and Terminal Classic period suggests salt works expanded in response to the increased demand for salt to meet basic dietary requirements of the larger population at inland cities in the Southern Maya Lowlands and to provide salted fish (McKillop 2005a). Salt continued to be made in southern Belize using the briquetage technique in the Postclassic at Wild Cane Cay and Frenchman's Cay (McKillop 2002:112) and in the Maya Highlands at Ixtápa and San Mateo Ixtatán (Andrews 1983:59, 85). Salt production continued throughout the Postclassic, Spanish conquest, and historic times, with Sacapulas, Ixtápa, and San Mateo Ixtatán still in production in the twentieth century.

In addition to the Paynes Creek Salt Works, salt works with briquetage are documented in lagoons along the coast of Belize at Placencia Lagoon; at Colson Point in central Belize; at Northern River Lagoon, Saktunja, Midwinter's Lagoon, and Salt Creek Lagoon north of Belize City; on Ambergris Caye; at Moho Cay, and 12 miles inland at Wits Cah Ak'al (Figure 1.1; Aimers et al. 2016; Andrews and Mock 2002; Graham 1989, 1994; Graham and Pendergast 1989; Guderjan 1988; MacKinnon and Kepecs 1989; Masson and Mock 2004; McKillop 2004; Murata 2011; Sills 2016, 2017; Valdez and Mock 1991). The briquetage technique also was used at inland salt springs, including Salinas de los Nueve Cerros and in the Maya Highlands in Guatemala and Chiapas, Mexico, at Sacapulas, San Mateo Ixtatán, Ixtápa, and many smaller salt springs (Andrews 1983; Dillon 1977; Dillon et al. 1988; Reina and Monaghan 1981; Woodfill et al. 2015). Elsewhere in Mesoamerica the technique was used along the Pacific coast of Guatemala and Chiapas, Mexico (Andrews 1983; Coe and Flannery 1967; Guernsey 2016; Nance 1992).

Brine-boiling vessels were variously supported by clay or stone cylinders above a fire or set on or into a stove or kiln. Ovens are described in France and China and may have been used in Vietnam and Guatemala at Salinas de los Nueve Cerros (Flad 2011; Olivier 2010; Reinecke 2010; Woodfill et al. 2015). Salt-enriched brine is boiled in shallow pans over a fire pit dug into the ground at Ban Non Wat, in northeast Thailand, and at the nearby Tung Pie Pong site (Halliwell et al. 2016:Figure 5.1; Nitta 1997). In Niger, pots are placed on cylinders in a kiln with a fire below (Li and Falkenhausen 2010:Plate 1.1–1.3).

In the Seille Valley in France, there are massive deposits of discarded remains of clay kilns and molds for salt cakes accumulated at salt works over a distance of 10 km (Olivier 2010; Olivier and Kovacik 2006). The salt works were located beside the salt sources where salty groundwater seeps to the surface from geological faults. At Marsal, one of the main salt works, hundreds of concentrations of kilns were set out in rows, with some measuring more than 100 m in length (Olivier 2010:Plate 6.1). Salt workers lived at adjacent settlements. Early Iron Age kilns held immovable brine-boiling shallow clay basins up to 50 cm in diameter, each with a capacity of about 30 liters of brine. Smaller, tapered clay containers dating later are interpreted as molds for forming and drying salt into salt cakes. The settlements and cemeteries beside the salt works included an impressive array of imported materials including a disc with coral inlay and gold leaf, gold earrings, and bronze jewelry, attesting to the high status of some of the residents.

Some salt works are known by the presence of solid-clay cylinders that were used as vessel supports during brine boiling. Solid-clay cylinders used as vessel supports are common at some sites in Vietnam, in England and elsewhere in Europe, and in Niger, Africa (de Brisay and Evans 1975; Gouletquer 1975; Reinecke 2010). The Gó Ô Chuà site in Vietnam is a 40,000 m^2 area containing an estimated 2–3 million clay cylinder "pedestals" (Reinecke 2010:141). There are no clay pots at the site, so Andreas Reinecke suggests that baskets or other containers of perishable materials were placed on top of the pedestals, which have three-pronged ends similar to those from Niger, where small bowls were placed on top of a single clay cylinder (Li and Falkenhausen 2010:Plates 1.1, 1.3). Dozens of cylinder-bowls were placed together and covered with a kiln, heated with firewood, vegetation, and palm fruits.

Clay cylinders are found at some modern and ancient salt works in the Maya area, notably Late Preclassic salt works at Ocós on the Pacific coast of Guatemala and along the coast of Belize at Northern River Lagoon, Moho Cay, Placencia Lagoon Salt Works, and the Paynes Creek Salt Works (Coe and Flannery 1967; Masson and Mock 2004; McKillop 2002, 2004; Sills 2016, 2017). A clay cylinder and small salt pot were in a burial at Moho Cay (McKillop 2004:Figure 16.5). In Europe and Vietnam, and ethnographically in Niger, the vessels were supported by a single clay cylinder with prongs or a concave surface to support the vessel (Li and Falkenhausen 2010:Color Plate 1.1–1.2). The clay cylinders vary in length, with extremely long clay cylinders used to support vessels over a fire in Niger (Gouletquer 1975). At the Paynes Creek Salt Works, clay cylinders are embedded at an angle in a base with a clay socket at the top with a concave surface on which the salt pot could rest. The longest clay cylinder at the Paynes Creek Salt Works is 33 cm (from Site 45), providing an indication of the height the vessel was supported over a fire. Most clay cylinders were fragmentary. Orientation of the cylinders at an angle in the base and the Maya practice of using tripod vessels suggest the salt pots were supported by three clay cylinders (McKillop 2002:Figures 3.1 and 3.15).

Studies of modern salt production indicate that apart from highly saline inland salt springs or wells, various techniques were used to concentrate the brine prior to evaporation over fires, including leaching, filtering, and burning salt-enriched plants (Andrews 1983; Antonites 2016; Good 1995; Gouletquer and Weller 2015; Halliwell et al. 2016; Reina and Monaghan 1981; Williams 2003; Yankowski 2010). The leached soil is a visible marker for salt production at many sites. Mounds of leached earth from brine enrichment are visible in the landscape in Thailand (Halliwell et al. 2016). Over 750 earthen mounds are scattered over the landscape of salt production at the Baleni salt works in South Africa (Antonites 2016:Figure 6).

Eduardo Williams (2003) describes solar evaporation in pans and leaching salt-saturated soil as two preferred methods to increase the salinity of the brine at modern sites in western Mexico. Both of those methods are reported in use recently on the Pacific coast of Guatemala. Such techniques were likely practiced as well in Mesoamerica in antiquity. Salinity increases seasonally during the dry season by solar evaporation in coastal lagoons along the north coast of the Yucatán and Belize. Inland salt spring and lakes vary in their salt content, depending on the underground salt source. Sodium (Na) and chloride (Cl) are up to 85 percent of inorganics in sea water, and inland brine sources are often higher

(Akridge 2008:1454). From ethnoarchaeological studies of salt making, Pierre Gouletquer and Olivier Weller (2015:14) found that "people don't try necessarily to get sodium chloride, but a 'generic' salt, the composition of which can be very different." San Mateo Ixtatán in highland Guatemala is described as having a high salinity in the salt spring, with 39 percent salinity (Andrews 1983:81–83). The brine from the salt spring was boiled without enrichment.

The Baleni salt makers collect salt-enriched soil that forms a hard crust around the edges of a marsh when the water recedes in the dry season (Antonites 2016:Figure 4). Baskets are used to drain brine after the salty soil is mixed with water (Antonites 2016:34–35). The brine is boiled, and moist salt is formed into salt cakes and allowed to dry further. The Ban Non Wat site and surrounding sites in the Khorat Plateau of northeast Thailand are located on an ancient seabed that has salt deposits 30–300 m underground (Halliwell et al. 2016). In the rainy season the land floods and draws salt to the surface. The salty soil is gathered in the dry season, January to April or May, when the water recedes and the ground dries. Salt is leached from the soil by soaking it in water in clay-lined basins (Halliwell et al. 2016:Figure 5.10). The brine is strained through filters and collected below. The brine is boiled in shallow metal pans over a fire in below-ground hearths (Halliwell et al. 2016:Figure 5.11).

A similar process of scraping salt-encrusted soil from a dry estuary in the dry season is practiced at the La Placita Salt Works in Michoacan, Mexico (Williams 2003). The salty soil is put into a raised, cone-shaped wooden structure. Salty water is poured on the soil and the enriched brine is filtered and collected below. In some cases, canals from the sea bring salt water to the salt works. The enriched brine is placed in salt pans for solar evaporation. The enriched brine is evaporated in 18 evaporation pans, each measuring about 6 × 3 m, with a capacity of 400 liters each. The salt is scooped out and dried in piles beside the salt pans.

Salt making on the island of Bohol in the Philippines uses organic material to absorb salt from sea water and then the brine created from pouring water on the burned ashes is boiled (Yankowski 2010:Color plates 8.1–8.2, 9.1–9.2). The process begins with making salt beds, which are square or rectangular pits in coastal mangroves measuring 7.2 × 7 m and 0.5–1.0 m deep. The salt beds are filled with coconut husks and left to soak in the salt water for two to six months. In areas without mangroves where the shore is open to the sea, coconut husks are weighted down by coral rock to soak in the sea. The coconut husks are then dried and burned, with the ashes placed in a cone-shaped filter made from bamboo. Sea water is poured in the filter, and the enriched brine is collected below in a wooden box. The brine is boiled in small open clay bowls set on a shallow metal pan heated by an enclosed oven below. As the brine evaporates to a wet salty mixture, it is transferred to clay jars, where it hardens. The salt is sold in the clay jars, with pieces of the base removed to use the salt. Since the salt is traded in the clay jars, the briquetage remaining at the site consists of the small brine-boiling bowls together with most other equipment consisting of perishable materials. Andrea Yankowski (2010:241) notes, "There are standard prices and trade equivalencies for salt, most of which are based on the size of locally-available stoneware jars."

Salt workers at Sacapulas sprinkle brine from a salt spring on sand nearby, let it dry, and repeat the process several times, finally collecting the salty soil (Reina and Monaghan 1981). It is carried in baskets and placed in a wooden box. Brine is poured through the soil, filtered, and collected below in clay jars. The brine is leached again until a small ball of cornmeal floats in the solution, indicating maximum salinity. The brine is carried in jars to the salt kitchen up the hill where it is stored until being boiled. About two dozen small open bowls are supported on stones above a fire. A solution of finely ground corn and water is added to the bowls to seal the interiors before brine boiling. Brine is continuously poured during the boiling process until the salt crystallizes. Loose salt is stored in jars. Alternatively, the bowls are turned over on the flat surface of the oven and hardened, requiring the bowls to be broken to remove the salt, leaving salt cakes for trade. The process is similar at Ixtápa, San Mateo Ixtatán, and other salt springs, except that the brine is collected from the salt spring and carried to the salt kitchens for boiling without leaching it through salty soil, due to the high salinity of the salt springs.

Ancient salt producers altered the taste, color, and texture of salt during production. Salt springs and coastal landscapes varied in the flavor of the salt produced (Aikman-Smith 2009). In central Mexico, the flavor was altered by filtering brine or fresh water through different soils (Parsons 2001). Black salt cakes produced at Sacapulas, Guatemala, were valued trade items (Reina and Monaghan 1981). The flavor of salt also was adjusted by controlling the evaporation process, since different salts precipitate out of salt water and crystallize at different rates (Bitterman 2010:38–39). During heating by solar evaporation or by use of fire, calcium salts precipitate first, then sodium chloride, then magnesium sulfate and potassium salts, followed by magnesium chlorine (referred to as bittern). For reference, modern table salt is generally pure sodium chloride. In Japan, artisan salt makers heat brine slowly to ensure that bittern, considered tasty, is included with the sodium chloride.

At many salt works, perishable materials often are used in the brine-boiling process, so there is little or no archaeological evidence. Salty water is poured through salty soil in old wooden canoes, large funnels, or wooden boxes, with the enriched brine collected below in clay pots or other containers. Where the salt makers also made the salt pots, the wooden pottery paddles (Halliwell et al. 2016:Figure 5.7) would not preserve. Wood containers were used for leaching salty soil or for boiling the brine in the Philippines and at the La Placita Salt Works in Michoacan, Mexico (Williams 2003; Yankowski 2010). Baskets are used to drain brine after the soil leaching in Thailand (Halliwell et al. 2016). Funnels made from bamboo are used to leach salty soil in the Philippines (Yankowski 2010). Salt kitchens are constructed of grass in Thailand (Halliwell et al. 2016). Salt kitchens are made from pole and thatch in the Philippines and Ixtápa in highland Chiapas, Mexico, and of stone at Sacapulas (Andrews 2003; Reina and Monaghan 1981; Yankowski 2010). A study of salt making in New Guinea left no durable evidence of salt production (Gouletquer and Weller 2015:16–18). The Dani in Western New Guinea collected specific plants as well as trees that produce little ash but big pieces of charcoal. The plants are soaked in a pond by a salt spring for about 1½ days and then dried and burned slowly, and the fire is dowsed with brine. The next day men pick out salt crystals from the ashes, mix the crystals with brine on leaves and press them into a mold. The salt cake is dried by a fire for over a week until it becomes as hard as rock. The gray salt cake is rich in sodium chloride, with few impurities.

Natural gas from brine wells fueled the extraction of brine in parts of China. Bamboo pipes transported brine for processing at sites in China. A 130 km long pipeline made from bamboo and raised on stilts transported brine from wells in the upper Yangzi basin to production salt works, attesting to the government's interest in having salt production nearby (Falkenhausen et al. 2006:335). At Lake Taihu near Shanghai, brine was conveyed in bamboo pipes to boats, each with a capacity of 1,000 liters, transported up a canal by buffalo, and then ladled into bamboo pipes to brine-boiling stations (Reinecke 2010:149). Transporting brine by pipe reached industrial production and distribution scale in England during the nineteenth century, where brine was piped to large metal containers inside production buildings (Hewitson 2015). Albeit on a smaller scale and not implying any governmental control, discovery of a pipe joiner at the David Westby Site (McKillop 2010a:Figure 4.6), suggests brine was piped at the Paynes Creek Salt Works.

In addition to use of natural gas in parts of China, wood was commonly used as fuel (Robinson and McKillop 2014), but peat also may have been used in England and elsewhere. Significant loss of heat from a fire is typical of the briquetage technique in which pots are raised above the fire to boil the brine, since heat is lost around the vessels (Akridge 2008:1457). Thermal conductivity of iron pans is higher than clay pots, which have thermal conductivity as low as 2–5 percent. Thin clay walls conduct heat better than thicker walls. Use of ovens or stoves is an increase in efficiency—less heat loss—compared to open fires, as in the Seille Valley. However, wood fuel for the kilns in the Seille Valley caused deforestation as well as sedimentation along the river valley (Olivier 2010: 257).

Briquetage at different sites on the coast of Belize includes two different styles, suggesting variation in the evaporation process. The use of open bowls, jars with a restricted orifice, or straight-walled basins supported by solid-clay cylinders, together with sockets, spacers, and bases, is common in southern Belize, including Paynes Creek and Placencia Lagoon, where the briquetage is designated "Punta Ycacos Unslipped" (McKillop 2002; Sills 2016; Sills and McKillop 2018a). The rims of the vessels are thick (as are the necks of jars) for holding or carrying, but the vessel bodies are thin, suitable for conducting heat. Solid-clay cylinders are common at Moho Cay, including one from a burial (McKillop 2002:Figure 3.52) along with a small salt-making jar. Solid-clay cylinders also are reported from coastal lagoon salt works in northern Belize (Mock 1994). Clay cylinders are known from Ocós and elsewhere on the Pacific coast (Coe and Flannery 1967). Stone vessel supports were used at Sacapulas.

The absence of clay cylinders from some salt works on the coast of Belize may indicate a different evaporation process, including the use of clay ovens or open-air evaporation in shallow pans. Clay cylinders are absent from Colson Point and rare or absent at Marco Gonzalez (Graham, personal communication 2010). Salt evaporation containers used at Colson Point and at Ambergris Caye and coastal lagoon in northern Belize include thin-walled vessels designated as the "Coconut Walk Unslipped" type (Aimers et al. 2016; Graham 1994). Described as platters, they resemble the vertical-walled basins of Punta Ycacos Unslipped pottery at the Paynes Creek Salt Works.

The Paynes Creek Briquetage

Artifacts on the sea floor at the Paynes Creek Salt Works provide information on the technology of salt production. The sample of artifacts in this study includes those recorded by site location during the 2003–2005 sea floor survey as well as artifacts that were individually mapped in the 2006–2008 sea floor survey. For the sites that were not piece-plotted during the 2003–2005 survey, I make comparisons among sites and quantify differences among sites in the briquetage. When the systematic survey began in 2004, the goals were to collect 30 measurable rims and 30 measurable vessel supports to look for variability among sites in the mass production of salt as measured by standardization of the briquetage, as previously done for three salt works, Stingray Lagoon, David Westby, and Orlando's (McKillop 2002:112).

Artifacts were collected during the 2006–2008 mapping survey to inform on the age of sites or identify distinctive activities associated with wooden buildings. Only selected artifacts were flagged, individually mapped, and surface collected. We flagged all obsidian and chert artifacts, ground stone artifacts, and wooden objects. We flagged all unit-stamped and other decorated pottery, Belize Red pottery, and any other unusual or distinctive pottery, such as fragments of large incense burners. Complete *candeleros*, figurine whistles, and formed clay objects such as boat models were flagged.

During mapping survey, only distinctive pieces of briquetage were collected as well as a sample of large rim sherds that showed the vessel shape. Some large rim sherds of bowls, jars, and vertical-wall basins and complete pots were flagged. Complete bases, sockets with or without clay cylinders, spacers, and complete clay cylinders were collected. Amorphous clay lumps, small sherds of various vessel shapes, and broken spacers and sockets were not recorded.

The pottery sample includes pottery flagged, mapped, and collected during underwater survey, which is a small percentage of the total pottery visible on the sea floor. Most of the pottery visible on the sea floor is briquetage, as quantified from subsequent excavations (McKillop and

Figure 3.1. Underwater view of pottery sherds on the seafloor at Site 49.

Sills 2016). The surface sample is biased in favor of pottery with painted, incised, stamped, or plastic decoration that we collected in order to help refine the temporal framework of the salt industry. Samples of briquetage, including complete vessels, large rim sherds of various vessel forms, and samples of vessel supports, sockets, spacers, and bases also were collected to look for variation in the briquetage.

The environmental context of the Paynes Creek pottery distinguishes the sample from pottery at other Maya sites. Since the artifacts were underwater, the pottery vessels, although broken, consisted of large sherds, which allowed collection of abundant information on vessel shapes and size and other characteristics. In a previous study of four salt works in the lagoon (McKillop 2002), I reported that pottery sherds were much larger and retained more of the vessel shape than sherds found at land sites in the area, making the study of pottery from the underwater sites worthwhile for estimating vessel shape and size. As shown in the underwater image of pottery sherds on the sea floor at Site 49 in the West Lagoon (Figure 3.1), the pottery from ongoing investigations is also large and well preserved, enabling estimates of the vessel shapes and sizes.

The salt water altered the surface color of the pottery, leaving sherds gray in color. The acidic mangrove peat dissolved limestone and dolomite temper in pottery, leaving voids that were visible as small holes on the surface of pottery (McKillop et al. 2018). The original color of painted pottery was rarely preserved. A glossy surface finish or decoration in shades of gray was occasionally preserved, especially when pottery was recovered from a matrix of thick silt, such as at the Stingray Lagoon site (McKillop 1995, 2002, 2011). When recovered from the underwater sites, the pottery was saturated with salt water. If allowed to dry, the salt comes to the surface and exfoliates the surface. Therefore, pottery was kept in water, with freshwater rinses for desalination. Each item was kept in a separate ziplock bag with its location marked on the bag. Once desalinated, pottery was allowed to dry, except for painted pottery which was kept in water to preserve the paint and surface finish. Selected pottery was stored underwater in silt at designated locations in Paynes Creek for long-term preservation.

Classification Methods

The ceramics were classified according to the type-variety system commonly used in Maya ceramic studies, using the types established for the coastal area (McKillop 2002). We also did an attribute analysis of the ceramics, describing 25 characteristics for each pottery sherd (including discrete observations such as shape as well as measurements such as rim diameter). Attributes of the pottery vessels such as lip treatment and decoration show patterns that may reflect temporal or spatial differences. The modal analysis is critical to identifying variation within types that may have temporal or spatial implications. Measurement data are used to discover differences in the standardization of the briquetage, using the diameter of the vessel opening of the salt pots and the diameter of the vessel supports, as in my previous study (McKillop 2002:112). A form was developed to record 25 characteristics of each pottery sherd. The data were entered into Excel spreadsheets to describe and analyze quantitative patterns and variability (standard deviation, mean, range, coefficient of variation, average median variation, and cross-tabulation). GIS was used to display the spatial distribution of pottery types within sites and across the lagoon system as well as the spatial patterning of different attributes, such as rim shape, motif, and type of vessel flange, which show functional and temporal differences. The GIS maps of K'ak' Naab' reveal briquetage outside of wooden structures but not inside, suggesting that the workshops were kept clean and workshop debris was discarded outside (McKillop 2007b). The chronological data from ceramic analyses, when added to the GIS, reveal there was a massive salt industry, with the salt works spanning the entire Late Classic period.

Digital and paper records of artifacts were sorted by site, with separate records for mapped artifacts and for those collected by site. Mapped artifacts were recorded during mapping by a unique letter, whereas artifacts collected by site were identified by a number corresponding to the number assigned when the artifact was drawn. Letters and numbers were specific to the sites from which the artifacts were found. For example, 33/180-2-1/1-A refers to Site 60 (33/180-2), sea floor (-1), pottery (/1), and artifact A. In contrast, 33/180-2-1/2-109 refers to Site 60 (33/180-2), sea floor (-1), chert (/2), and artifact 109 (from the drawing #).

For the mapped artifacts, descriptive information was added to an Excel file with the Universal Transverse Mercator (UTM) locations of each artifact. The data were copied into a single Excel file, which was then attached to the project Underwater Maya GIS. Data updated in the Excel file was automatically available in the GIS. The artifact data, either complete or in part, could be viewed alone or along with the wooden post data. Numbered artifacts from the 2003–2005 survey, whose location was recorded by site but not individually mapped, were assigned UTM locations near the site. The numbered and lettered artifacts were distinguished in the GIS by different symbols.

A total of 253 pottery sherds and other items of Punta Ycacos Unslipped were individually mapped in addition to sea floor collections made in 2003–2005. Samples of briquetage including complete vessels, large rim sherds of various vessel forms, and samples of vessel supports, sockets, spacers, and bases were collected to look for variation in the briquetage (Tables 3.1–3.3). Amorphous clay lumps, small sherds of various vessel shapes, and broken spacers and sockets were not recorded. Other fired-clay objects made from the same clay and quartz sand temper, assigned to the Punta Ycacos Unslipped type (despite not being from pottery vessels), include figurines, clay boat models, *candeleros*, large incense burners, and a pipe reducer.

Briquetage includes sherds from jars, basins, and bowls as well as solid-clay cylinder supports used to hold the vessels over a fire during brine evaporation (Figure 3.2, Tables 3.1–3.3; McKillop 1995, 2002, 2005a). Vessels used to evaporate brine include jars, bowls, and basins with thin

Figure 3.2. Briquetage from a 10 cm level in a 1 × 1 m unit in Transect 1 at Site 29.

walls that would have maximized heat conductivity. The rims are thick, a trait useful for holding or moving the containers. The vessel interiors are smooth, which is useful when scraping out the crystallized salt. The vessel exteriors vary but are rough, which would have aided in holding the containers. Jars include necked jars and jars with outward-curved walls and with straight or outward-folded lips (Figure 3.3). Basins are large,

Table 3.1. Number and percentages of different salt-making vessel shapes (Punta Ycacos Unslipped) from the mapped artifacts based on rim sherds

Shape	Number	Percent
Bowl	31	24.0
Dish	4	3.1
Jar	43	33.3
Plate	2	1.6
Tecomate	1	0.8
Basin	48	37.2
Totals	129	100

Table 3.2. Summary diameter data from the mapped Punta Ycacos Unslipped rim sherds

Shape	Number	# Measurable	Mean diameter	Maximum diameter	Minimum diameter	Median diameter
Bowl	30	26	29.1	46	7	31.5
Jar	45	38	22	44	5	17
Basin	48	42	23	35	8	26
Dish	4	4	32	46	23	—
Funnel	21	13	3.2	4.6	2.5	—
Incense burner	29	7	23	28	13	—
Candelero	13	6	1.7	2	1.5	—

Table 3.3. Numbers and percentages of mapped nonvessel briquetage (Punta Ycacos Unslipped)

Shape	Number	Percent
Socket	9	26.5
Spacer	6	17.6
Cylinder	11	32.3
Base	6	17.6
Spout	1	3.0
Pipe reducer	1	3.0
Totals	34	100

straight-walled vessels, usually with an outward-folded lip. There are bowls of various sizes and shapes, most of them round-sided and open.

The assemblage of briquetage used at the Paynes Creek Salt Works included vessels supported by clay cylinder vessel supports attached to a base and to a socket at the top, on which the vessel rested, as well as clay spacers that may have been used to separate pots from one another over a fire. Flat bases with cylinders embedded at an angle indicate the vessel supports were slanted (McKillop 2002:Figure 3.15). There likely were three cylinders per vessel, following the Late Classic tripod tradition. The top end of each cylinder had a socket with a concave upper surface on which the vessel rested. Multiple salt pots placed over a fire may have been separated from each other by spacers, which are disks with concave surfaces (McKillop 2002:Figures 3.19–3.20). Also recovered were quantities of amorphous clay lumps, baked clay items of various shapes and sizes

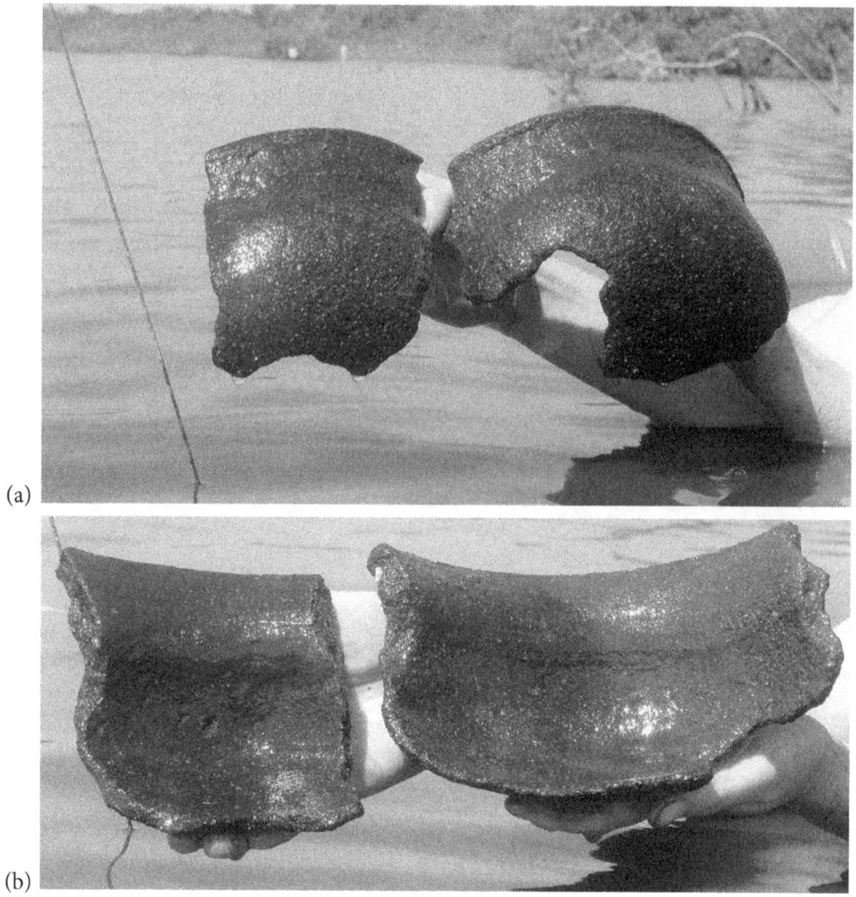

Figure 3.3. Punta Ycacos jars from Site 56 (Stingray Lagoon site), showing (*a*) exterior and (*b*) interior views.

that are the residue of pottery manufacture, as well as discarded pots and other briquetage no longer useful for brine evaporation. Hearths consisting of congealed clay, charcoal, and briquetage were excavated at the Stingray Lagoon site and in the Main Channel at Mudshell and Site 100 (McKillop 2002:38,2017a). The sockets, spacers, and solid-clay cylinders were made from poorly mixed clays and fired at low temperatures. The length of complete cylinders provides an indication of the height that pots were placed over a fire during the evaporation process: Complete cylinders ranged from 29.5 cm at the Stingray Lagoon site (McKillop 1995:Figure 11) to 33 cm at Site 45. Most cylinders were broken and discarded after use. Some cylinders became amorphous clay lumps. Some of the clay lumps were likely left over in the production of vessels and other parts of the briquetage assemblage.

Fuel has been described as a limiting factor in salt production where brine is evaporated by heating in pots over fires (Early 1993; Mock 1994). This was not a problem at some salt works in China associated with natural gas wells (Flad 2011; Flad et al. 2005). Where wood was used for fuel, historically and ethnographically around the world, techniques were used to concentrate the salinity in the brine before it was heated, thereby reducing the amount of fuel needed (Adshead 1992). Leaching brine through salty soil is a common method used to obtain a brine of high salinity. Salt water is poured through salty soil held in raised wooden platforms, funnels, boxes, and old canoes, with a container below to receive the brine (Good 1995; Halliwell et al. 2016; Reina and Monaghan 1981; Williams 2003). In the Philippines, wood and coconut husks were chopped into small pieces, soaked in salty water, dried, and then burned to ash, with the salt leached and then evaporated in clay pots to make salt (Yankowski 2010). In some cases the flavor of salt is controlled by pouring fresh water through different types of salty soil, as described for Nexquipayac in central Mexico (Parsons 2001).

Earthen mounds along the shore of Placencia Lagoon described as the remains of the brine-leaching process (Sills 2016, 2017) underscore the likely former ubiquity of leaching mounds at the Paynes Creek Salt Works (McKillop 2002:49; Watson and McKillop 2019; Watson et al. 2013). An earthen mound at the Killer Bee salt works in Paynes Creek National Park is interpreted as a slag heap remaining after brine had been leached through salt-saturated soil. Two earthen mounds at Witz Naab' also are slag heaps from brine enrichment. At the underwater salt works elsewhere in Paynes Creek, any similar heaps would have been submerged below the sea.

Due to the ubiquity of red mangrove peat and the shortage of soil in the immediate area of the Paynes Creek Salt Works, salty peat may have been used to enrich the salt content of brine before it was boiled in the salt kitchens. The salt works are located in a mangrove ecosystem dominated by red (*Rhizophora mangle*), black (*Avicennia germinans*), and white mangroves (*Laguncularia racemosa*), which also were characteristic of the landscape when the salt works were in production (Jones 2013; McKee 2011; McKillop and Jones 2018; McKillop et al. 2008, 2010a, 2010b). Red mangrove peat extends from the sea floor to depths over 4 m in the lagoon

(a)

(b)

Figure 3.4. Punta Ycacos funnel from Carpenter site (61W), (*a*) exterior and (*b*) interior views.

system at the Paynes Creek Salt Works. The peat formed as a result of red mangroves growing taller to keep their leaves above water as sea levels rose (McKillop 2002:135–174). The peat is easy to extract from the lagoon floor by cutting with a sharp implement. The peat retains its shape and may have been dried in the plazas and stored indoors. With a high organic content, as calculated by loss-on ignition values (McKillop et al. 2010a, 2010b), the signature of chemical testing of the salt kitchens and yards would be expected to be high in phosphorus and sulphur. A low-lying area at the Chan b'i site was interpreted as a possible extraction location for leaching sediment (Sills et al. 2016).

Some artifacts may inform on the technology of production, including the presence of funnels, vessels with a small hole, the pipe reducer, and manos and metates. Funnels made from sand-tempered clay were common at the Paynes Creek Salt Works, indicating that brine enrichment was commonly practiced (Figure 3.4). Earth mounds representing soil depleted of salt content from the brine-enrichment process may have been more ubiquitous on the landscape as at the Placencia Lagoon Salt Works (Sills 2016, 2017), but sea-level rise and wave action washed away any earth mounds that may have been present at individual salt works. The occurrence of clumps of red mangroves beside many of the sites in the East and West Lagoons may be relict brine-enrichment mounds. The mangroves may have taken root on earth mounds as sea level rose in antiquity.

A thick-walled clay funnel was found under the wooden canoe at Site 67. The canoe was supported by wooden stakes and located between two lines of palmetto palm posts. By ethnographic analogy with modern salt production, the canoe was filled with salty soil and then fresh or salty water was poured into the soil-laden canoe, which had holes in the base, through which brine drained into the funnel and then into water jars for storing the brine before it was evaporated over fires (Andrews 1983; Williams 2003). Jeffrey Parsons (2001) reports that strainers made from vegetation were used to filter sediment, which must have occurred at the Paynes Creek Salt Works as well. Filters may have been placed at the base of the canoe and in the base of the funnel. The widespread occurrence of Punta Ycacos sherds with small holes suggest they functioned as funnels, similar to the funnel found under the canoe at Site 67, the Eleanor Betty site (McKillop et al. 2014).

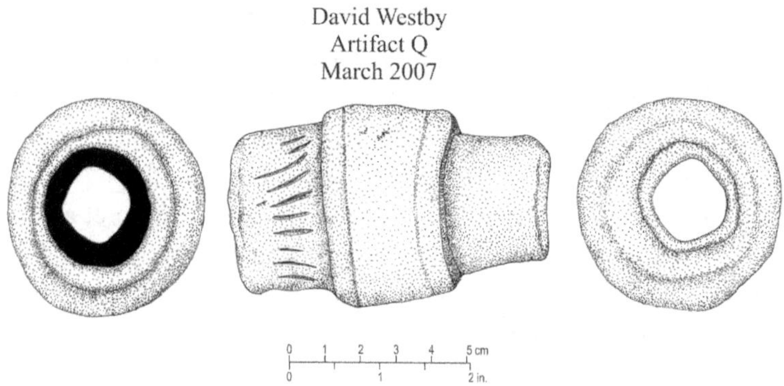

Figure 3.5. Pipe reducer from the David Westby site (Site 57), showing profile and end views. By Mary Lee Eggart.

The recovery of a clay item identified as a pipe reducer (Figure 3.5), used to join two pipes of different size openings, suggests the use of a piping system for brine as seen in Late Medieval China (Falkenhausen et al. 2006:335). The pipe reducer may have been used in conjunction with large, thick-walled clay vessels with a hole at one end, as have been found at many sites. A possible relict salt pan observed from remotely piloted aircraft air photo coverage of the lagoon may have been used to enrich the brine, which was then piped to the nearby Orlando's and David Westby sites for evaporation in pots over fires.

At the Paynes Creek Salt Works, the linear arrangement of palmetto palm posts that demarcate plazas or yards beside many of the salt kitchens suggests that most of the palmetto palms were retaining walls used to shore up the land at the edge of salt works to prevent erosion and encroachment of rising seas. Some of the linear alignments of palmetto palm posts may have defined salt pans. They may have provided an additional mechanism for increasing the salinity of brine before the boiling process or, in some cases (as at La Placita), instead of brine boiling. A rectilinear raised feature observed from the remotely piloted aircraft air photo near Site 58 may be a relict salt pan. It resembles relict salt pans marked by lines of living black mangroves (*Avicennia germinans*) on the shores of the Gulf of Nicoya in Pacific Costa Rica (Bishop 1977:Figure 39).

Punta Ycacos Unslipped also includes *candeleros* and large incense burners that were commonly found on survey but only selectively mapped since they are not temporally diagnostic (Table 3.2). The *candeleros* include tripod and flat-bottomed shapes. Large incense burners include lids with modeled forms of lizards, pedestal bases, vertical-wall vessels with spikes or ridges, and a prong from an incense burner.

Summary

Abundant briquetage from mapped artifacts found in flotation survey of the lagoon system indicates that salt was produced by evaporation in pots over fires inside wooden buildings. Pots include jars, bowls, and vertical-wall basins. They were smooth on the inside and rough on the outside, with thick rims and thin bodies. Each pot was supported by three solid-clay cylinder vessel supports set into wet clay at an angle. The length of complete cylinders indicates the height that the vessels were raised over the fire, with longest cylinders measuring 29–36 cm. Most cylinders were fragmentary, including some with attached bases or sockets, the latter with a concave surface where the pot rested. Spacers with two concave surfaces were used to separate jars over fires during evaporation.

Charcoal was abundant, including hearths excavated at Sites 56, 100, and Mudshell (McKillop 1995, 2017a). To reduce the wood fuel needs for evaporation, the salt content of brine was enriched by leaching salty water through salty soil or by solar evaporation. Both techniques were used at the Paynes Creek Salt Works. Chemical testing of sediment along transects at Site 24 using inductively coupled plasma-mass spectroscopy indicates areas for hearths, disposal of briquetage, and remnant brine-enrichment discard (Sills et al. 2016). The salt content of brine was increased by pouring salty water through salty soil in large containers such as the wooden canoe at Site 67. The Site 67 canoe was supported by short wooden stakes and had a large pottery funnel underneath the canoe to channel enriched brine into a pot. Funnels were commonly recovered during flotation survey.

Earthen mounds were excavated at Killer Bee and Witz Naab', the only visible markers on the landscape of the brine-enrichment process (McKillop 2002; Watson and McKillop 2019; Watson et al. 2013). Leached mounds of soil were more common but were deflated by wave action after the salt works were submerged by sea-level rise. Patches of red mangroves beside most underwater salt kitchens may be relicts of leached mounds.

Solar evaporation may also have been used to enrich the brine. A rectilinear feature observed on a low-level image of the lagoon may be a relict salt pan. Brine may have been transported from the salt pan to salt kitchens, as suggested by recovery of a pipe joiner at Site 57. Brine was piped in China (Falkenhausen et al. 2006:335).

The briquetage recovered from the Paynes Creek Salt Works fits the ethnographic model at Sacapulas in terms of enriching the brine by pouring it through salty soil and then boiling the brine in pots over a fire inside a salt kitchen. A wooden box was filled with salty soil from the playa beside the salt spring at Sacapulas. A wooden canoe was used at least at one of the Paynes Creek Salt Works (McKillop et al. 2014). At Sacapulas, the enriched brine was collected below the wooden box in large clay pots. A funnel was found under the raised canoe at the Eleanor Betty Site. At Sacapulas, as at other salt works where salty water is leached through salty soil, the enriched brine was strained through filters consisting of grasses or other organic matter before it was collected in a container. A ceramic disk with many small holes from the Paynes Creek Salt Works resembles an object from a salt work in the Manga region of Niger (Li and Falkenhausen 2010:Plate 3.1) described as a strainer. The strainer may have been placed over the hole in the bottom of the canoe, and vegetation may have been placed on the strainer to filter out the soil. About two dozen small, open bowls were elevated above a fire inside a salt kitchen at Sacapulas. Although stones were used to raise the bowls, solid-clay cylinders were used at the Paynes Creek Salt Works. In addition, several different vessel forms were used, including jars, bowls, and basins. Large jars were used at Sacapulas to store brine and loose salt inside the salt kitchens. Large Mangrove Unslipped jars as well as smaller Warrie Red jars were used at the Paynes Creek Salt Works.

Surplus production of salt changed the value of salt from a basic necessity to a unit (salt cake) for trade, an item of value that could be stored (salt in the bank) for later trade and used to make other commodities that could be stored or traded, notably salted fish or fish sauce—a widely traded commodity in ancient Roman times, in China, and elsewhere (Yankowski et al. 2015). Gouletquer and Weller (2015:24) describe salt as a "pre-monetary wealth, able to be used as an ordinary food for men and cattle, used for trade and barter, offer as a gift or hoarded. That is why one can call it an 'ingot.'"

Ethnographic and ethnoarchaeological studies of salt production are insightful in providing models for ancient salt making due to the diversity of techniques and the differences in preservation at ancient sites. The preservation of wooden salt kitchens, wooden objects, calabash bowls, and the large size of the briquetage allow a rich reconstruction of salt production. The size and shapes of briquetage vessels can be reconstructed, whereas this is not always possible at land sites, where briquetage vessel shapes are difficult to discern. The preservation of salt kitchens allows the spatial analysis of salt production, contingent activities, and possible residential use. The preservation of wood allows identification of species, which allows wood selection strategies to be discovered, patterns of wood species selection by salt workers at adjacent salt kitchens to be evaluated in terms of surplus household production or community production, the diameters of wooden posts to compare with traditional Maya pole-and-thatch buildings in more recent times, and evaluation of wood chipping of the ends of posts driven into the ground. The mangrove peat holds the salt kitchens and associated materials in place against the ravages of hurricanes, under the shallow waters of the coastal lagoon, protected from postdepositional trampling by people and animals.

4

The Infrastructure of Salt Production

Wooden Architecture

The Paynes Creek wooden buildings underscore the information about ancient cultures that can be recovered from peat bogs and other environmental settings that preserve wood from the distant past. Preservation of wood at the Paynes Creek Salt Works indicates there was an infrastructure of production and distribution of salt that was carried out inside wooden buildings and distributed by canoe transport (McKillop 2005a). Worldwide, ancient wooden buildings are rare recoveries that are preserved in frozen ground, permanently dry conditions such as deserts, or waterlogged environments (Coles and Coles 1989; Purdy 1981, 2001). Wooden posts, beams, and rafters as well as palm-thatched roofs were preserved in the shallow coastal waters of Los Buchillones, Cuba (Pendergast et al. 2001). Wooden structures are preserved in peat bogs at the Must Farm in England and various other sites in Europe, and in mudslides on the west coast of Canada and the United States, notably at the Ozette site in Washington State.

The range of organic remains variously preserved in these settings includes wooden structures, canoes, and other wooden artifacts; basketry; cordage; cloth; and human bodies. Burials, including of brains and other flesh, were discovered at the Windover site in Florida in a shallow pond (Doran 2002). Many preserved human bodies have been recovered from peat bogs in England, Scotland, and northern Europe, where cool temperatures preserve the bodies. Peat bogs with low pH, such as the Ozette site, tend not to preserve bone (Croes 1976). The warm temperature of the Belize mangrove peat, together with the acidic sediment (McKillop 2005a), indicate that although wooden structures and stone and pottery artifacts are ubiquitous, the setting is not conducive to preservation of human bodies or bone.

The Paynes Creek Salt Works provide a model for other salt works where brine is evaporated in pots over fires along the coast of Belize and the Pacific coast of Guatemala but where wooden buildings have not preserved. Although there are ethnographic examples of salt produced outside, such as in western Mexico and in the case of solar evaporation ponds (Andrews 1983; Williams 2003), indoor salt production provides a place to store equipment, supplies, and salt as well as protection from rain. For example, the salt kitchens at Sacapulas are used to store firewood, loose salt in large pots, and jars with brine. Brine boiling takes place on a fire hearth in a raised platform in the middle of the building (Reina and Monaghan 1981).

Indoor production means salt can be produced without concern for rain. Even the dry season in southern Belize, between March and mid-June, can include significant amounts of rain. A 30-year rainfall record on the coast near the Paynes Creek Salt Works indicates variability in monthly and yearly rainfall amounts, with rain occurring in the driest months (Figure 4.1).[1] The use of buildings also extends salt production beyond the dry season since brine and even salty soil can be stored indoors. Salty soil is banked inside salt kitchens at Sacapulas during the dry season so that salt can be made during the rainy season if there is a demand (Reina and Monaghan 1981:23). The potential for year-round salt production is in line with market demand for salt, which is not limited to the dry season. Certainly, salt can be stored, but shortages can occur, so the capability of making salt on demand, off season, is a good marketing strategy.

There are several possible explanations for the current underwater setting of the Paynes Creek Salt Works. They may have been submerged by actual sea-level rise, relative sea-level rise (subsidence), tectonic activity, or a combination of these. If the Classic Maya cleared mangroves to build salt kitchens, the clearing would have made the land subside. Sea-level rise has been documented by the occurrence and depth of red mangrove peat below the sea floor in the lagoon since the peat is deposited as red mangroves rise to keep their leaves above water (McKillop et al. 2010a, 2010b). An earthquake May 29, 2009, that measured 7.1 on the Richter scale collapsed posts five feet into the ground that supported wood-frame

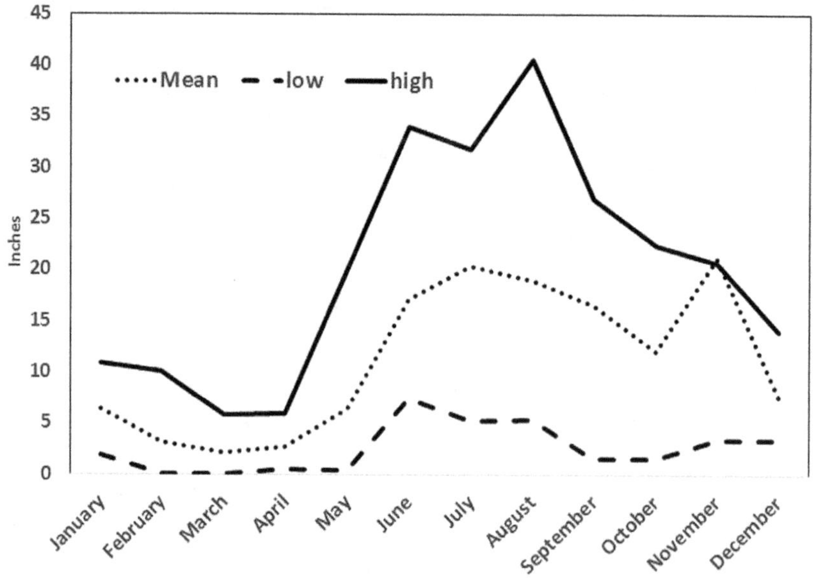

Figure 4.1. Monthly rainfall at Village Farm, Toledo, 1980–2005, from daily rainfall records by Tanya Russ and John Spang.

buildings elevated on posts in Monkey River Town to the north of Paynes Creek National Park (Parks 2009). Alternatively, the Paynes Creek sites could be the collapsed remnants of buildings constructed on platforms over the sea.

Water depth at different sites varied from a few centimeters to over a meter, with the tops of posts, pottery, and stone tools visible on the sea floor, although often obscured by a lens of loose silt. The silt layer was generally less than 10 cm thick, except in the Main Channel, where the silt was up to 1 m in thickness. Thick silt of about 20 cm preserved the surface of painted pottery and preserved ancient botanical material, including palm frond stems, mamey apple seeds, and calabash at the Stingray Lagoon site, which is in the middle of the Stingray Lagoon area, 300 m from the nearest shore. Posts extending above the sea floor were worm-eaten and decayed. In contrast, the posts were perfectly preserved—albeit salt-waterlogged—in the mangrove peat sediment below the sea floor. Mangrove peat also covered many posts, partly obscuring Sites 60 (Ek Way Nal) and 72 (the John Spang Site) and likely completely hiding some sites. Living mangroves and mangrove peat have grown over some sites, notably Site 60, which extends from the Main Channel underneath a bar of mangroves into the East Lagoon. Shovel tests in the black mangroves at the Killer Bee site revealed archaeological deposits to 80 cm depth. Artifacts were visible on the black mangrove mud flats around the Witz Naab' mounds (Site 50) as well, suggesting that the mounds at both sites are the visible remains of larger salt-production sites. The depth of the archaeological deposits below the sea floor and the depths of the ancient building floors were unknown until later excavations of selected salt works (McKillop and Sills 2016; Sills and McKillop 2013). Sediment sampling using a T-core to extract samples below the sea floor in 2004 reached the maximum depth capability of the corer at 4.3 m, with mangrove peat the entire depth (McKillop 2005a).

A Model of Sea-Level Rise and Salt Production

The production of salt inside wooden buildings at the Paynes Creek Salt Works provides a model for salt production along the coast of Belize where briquetage is present but wooden buildings have not preserved. Previous investigation of four salt works in Paynes Creek suggested that salt was produced by people who lived elsewhere, was specialized in production as indicated by the standardization of the salt pots, and was produced for consumers at the nearby inland Maya at cities in Belize as indicated by the presence of inland trade goods at the salt works (McKillop 1995, 2002).

Discovery of the wooden architecture in 2004 and subsequent investigation of 110 sites at the Paynes Creek Salt Works focused on investigating whether salt making was a household activity, whether there were residences at the salt works, and whether there was evidence of state management at the salt works, either by direct control by a state overseer or by evidence of surplus production for trade or tribute such as warehouses or other storage areas. Radiocarbon dating of wooden posts and modal analysis of stylistic differences in nonbriquetage pottery were used to develop a chronology to see if all sites were used at the same time, indicating a massive industry, or, alternatively, if one or a few salt works were used at a time over several hundred years, indicating small-scale production, likely surplus household production by families. In addition, some salt works may have been used for a long time. The abundance of briquetage indicates the sites were used in the production, storage, and distribution of salt. Salt was produced by evaporating brine in pots over fires. This activity was indoors, as revealed by abundant broken salt-making pottery vessels inside structures. Salt was in demand at Classic period inland cities where this biological necessity was scarce and where salt preservation of fish and meat would have been necessary.

The mangrove peat that preserved the Paynes Creek wooden buildings consists of a highly organic marine sediment that accumulates as red mangroves (*Rhizophora mangle*) keep pace with sea-level rise in tropical areas. The sediment consists of up to 75 percent organic matter (McKillop et al. 2010a, 2010b). Sorting the organic matter under a microscope revealed the organic matter was mainly fine red mangrove roots. If seas rise slowly, as they did at the end of the Pleistocene, red mangroves grow taller to keep their leaves above water. If seas rise quickly, the mangroves are drowned. As sea level rose in the Holocene along the coast of Belize, the area between the coast and present barrier reef system flooded, ultimately creating the inshore barrier reef lagoon, with some 200 islands, leaving the barrier reef up to 40 km offshore in southern Belize. Sediment cores yielded as much as 11 m of mangrove peat below the sea floor in the barrier reef lagoon in southern Belize (Macintyre et al. 1995; McKillop 2002). Red mangroves subsequently took hold in areas of shallow water, creating hundreds of offshore cays. Many of the mangrove cays are aligned with submarine limestone ridges that parallel the coastline and may represent ancient shorelines.

A model of the settlement history of the Paynes Creek Salt Works includes lower sea levels that exposed large areas of the sea floor in Punta Ycacos Lagoon before the beginning of the Classic period. The first wooden salt kitchens were constructed near the shorelines on solid ground composed of red mangrove peat. Microtidal variations of 0.5 m indicate that land may not have been much above sea level. The sea floor is periodically exposed above water in shallow-water areas of Punta Ycacos Lagoon during a "norther" wind that keeps the tide out for several days, resulting in shallow-water salt works kept exposed above water.

The salt works were located along the shores of Punta Ycacos Lagoon where solar evaporation in the dry season meant that the salt water lagoon was more saline than the open waters of the adjacent Caribbean. Using the hypersaline lagoon water reduced the amount of wood fuel needed to evaporate brine in pots over fires. Since the dry season is short and rainy, collecting salty water from the lagoon, enriching the salt content by pouring the brine through salty soil, and storing the brine in large water jars inside wooden buildings extended the salt-production season compared to production outside.

Salt production in pole-and-thatch buildings along the shores of Paynes Creek for producing salt and storing brine, loose salt, and salt cakes was similar to the process at Sacapulas (Reina and Monaghan 1981). The Paynes Creek Maya collected lagoon water and enriched its salt content by pouring the lagoon water through salty soil in old canoes or other wooden containers. Enriching the salinity of the brine reduced the fuel costs for brine boiling. The salt-enriched brine was strained through leaves or woven mats placed over a hole in the bottom of the wooden container. The salt-enriched brine was collected in pots underneath, stored in large jars inside the salt shed, and ultimately evaporated in small bowls or jars over a slow fire, producing salt. Salt may have been transported in jars or bowls, as reported in an ethnoarchaeological case study of brine boiling in the Philippines. In that study, Andrea Yankowski (2010) reports that the base of each pot was broken to expose the salt for use. The pots were wrapped in leaves for transport. Alternatively, the salt cakes may have been removed from the pots before transport as at Sacapulas, where bowls of hardened salt were turned upside down over a fire to further solidify the salt. Since the salt adhered to the bowl walls, the pots were broken to reveal the salt cakes.

The process of salt production described at Sacapulas is common historically and in modern times worldwide, including enrichment of the salty water before it is evaporated over fires, in prehistoric France, North America, the Philippines, and Thailand (Brown 1980; Halliwell et al. 2016; Olivier 2010; Yankowski 2010; Yankowski and Kerdsap 2013; Yankowski et al. 2015). Brine from wells near Zhongba in China and at Ixtápa and San Mateo Ixtatán in the Maya highlands was high in salinity and was therefore boiled in pots without further enrichment of the salt content (Andrews 1983; Flad 2011:40). Eduardo Williams (2003) describes canals dug to bring salt water from the sea, which was poured through salty soil in elevated platforms. Pots were placed underneath to catch the saltier water, which was then evaporated in small clay evaporation ponds. The archaeological correlates of this process include briquetage—the pots used in the evaporation process, mounds of salt-depleted soil from the brine-enrichment process, and, in some cases, small solar evaporation ponds as well as canals or pipes, as in ancient China to transport brine to the salt works (Falkenhausen et al. 2006:335; Halliwell et al. 2016; Lin 1940; Williams 2003).

Surplus Household Production in the Maya Economy

The Maya household was the basic unit of the subsistence economy, in which resources were extracted and food and craft items prepared (Hirth 2009a, 2009b; Robin 2012; Wilk 1988). The domestic economy of household production includes activities within and around the domicile itself as well as activities carried out by householders at other locations, such as farming, fishing, and hunting, which are located near resources, some of which are seasonally exploited at sea or inland at farms distant from homes, as articulated by Bernard Nietschmann (1973) for coastal settlers on the Caribbean coast of Nicaragua. Craft production was carried out by households for their use and for distribution elsewhere within the community and even beyond. Some products had regional distribution, notably the high-quality chert objects produced in household workshops at Colha in northern Belize (Shafer and Hester 1983).

The household is reflected architecturally by a range of buildings, from isolated structures to clustered groups of buildings, including plaza groups where buildings were arranged to enclose a central open space (Robin 2012). Ethnographic studies of modern Maya villages underscore variability in relationships among people that underlie household composition and postmarital residence patterns, which are important variables in interpreting the use of buildings at ancient Maya sites (Wilk 1988). Modern and historic Maya households include isolated nuclear-family households as well as multiple-family households. Multiple-family households in the Maya area can occupy a single large structure or several structures, so there is not an equivalency between house and family archaeologically. Multiple-family households are common among the Chorti, Mopan, Ketchi, and Yucatec Maya (Redfield and Villa Rojas 1962; Wilk 1988).

Examining case studies of salt production from around the world reveals variability in the location and seasonality of salt workers' residences. Salt makers often live in residences near salt works, either in the same community or nearby, or they migrate seasonally to salt works. At Sacapulas, the salt makers' residences are located behind the salt work buildings, which were adjacent to the salt spring (Reina and Monaghan 1981).

Rowan Flad (2011) suggests that the salt workers at Zhongba on the Yellow River in China lived nearby but not at the salt works. Salt works identified by briquetage on the southeast coast of China in Shandong Province had worker houses farther back (Flad 2013). Jeffrey Parsons (2001) notes that salt workers lived in Nexquipayac approximately two kilometers from their salt works. Wild Cane Cay would have been easily accessed by a short canoe paddle of 7–8 km to the Paynes Creek Salt Works, and there are other nearby coastal settlements (see Figure 1.2). Salt production can be year-round or seasonal. In western Mexico, inland settlers visited the coastal salt works seasonally and set up temporary camps nearby (Williams 2003).

If the salt workers lived at the Paynes Creek Salt Works, either in separate houses (as at Sacapulas) or in the same building or plaza group, we would expect to find evidence of household activities such as those found in Maya residences at nearby Wild Cane Cay or at Chan in western Belize (see McKillop 2005b; Robin 2012). Material remains of household activities can include middens with plant and animal food remains; burials associated with houses; a variety of pottery vessel forms for household use in storage, cooking, and serving; and obsidian, chert, and ground stone tools. We would expect multicrafting, including remains of salt production and domestic activities.

If evidence of residential activity is lacking, then the salt workers lived farther away—either on the coast at Wild Cane Cay or another nearby coastal settlement, at least during salt production. Concentration of the salt works in the lagoon areas close to the sea instead of farther away in the lagoon support the idea that the salt works were accessed from the sea rather than by walking overland. Workers from the coast or inland communities may have visited the Paynes Creek Salt Works seasonally and set up temporary camps, as reported by Williams (2003) for western Mexico. Activities at nonresidential salt works may have been solely focused on salt production or may have included related activities such as cleaning and salting fish.

Coastal salt production for inland transportation leaves archaeological evidence. If salt was traded inland, there would be evidence of inland trade goods at the Paynes Creek Salt Works. In some cases salt was transported in clay pots from coastal salt works to inland consumers, which would skew the evidence of briquetage at the salt works since most pots were transported with salt, and few pots remained at coastal sites (Yankowski

2010). If the salt pots were wrapped in leaves for transport as in the Philippines, evidence at inland marketplaces would include salt pots as well as organic residue from the leaves and from any other material that adhered to the leaves during wrapping and transport. However, if salt cakes were transported without pots from the Paynes Creek Salt Works, there would be no material evidence at inland consumer sites. If the salt works were directly controlled by officials of inland dynastic polities, there would be architectural remains of their residences and offices. State control of salt production in ancient China provides an analogy (Flad 2011). The coastal salt works in Shandong Province had overseer quarters even farther inland from the salt workers' residences (Flad 2013). If salt was commissioned by inland dynastic Maya, alliances including gifting of status items to coastal elite may have developed to foster and maintain a regular supply of this basic biological necessity to salt-starved Maya at inland cities. The depiction of a "salt person" painted on the exterior of a building at Calakmul underscores the interest in salt by the inland, urban Maya during the Classic period (Carrasco et al. 2009). The salt person and images of other producers/traders/vendors are located near the marketplace.

Previous Evidence of Ancient Maya Wooden Buildings

In his settlement study at Tikal, William Haviland (1985) notes that wooden structures for houses, workshops, storage, and other uses were likely common at the ancient city, despite their lack of preservation. Stone temples, palaces, and other elite architecture dominate the modern landscape of ancient Maya sites (Abrams 1994; Christie 2003; Ferguson and Adams 2001; Harrison 1999; Miller 1999; Reents-Budet 2001). Pictorial depictions on painted pottery vessels and murals for the most part focus on activities of the royal court associated with stone buildings. The painted pots show scenes of Classic Maya royalty receiving visitors, tribute, war captives or feasting, fasting, or engaging in bloodletting or other vision quests, often indoors (Coe and Kerr 1998; Miller 1999; Reents-Budet 1994).

Wooden buildings made of perishable materials were prevalent in antiquity, as they are today in traditional Maya communities (Figure 4.2). Depictions of pole-and-thatch structures, such as the mural scenes on the Temple of the Warriors at Chichén Itzá, are less common than depictions of stone architecture. Pole-and-thatch structures that leave a visible

(a)

(b)

Figure 4.2. Traditional Maya pole-and-thatch buildings, (*a*) under construction and (*b*) completed. Photos courtesy of John Anderson, the Robert C. West Latin American Photographic Collection, LSU Cartographic Information Center.

mark on the modern landscape include groups of buildings that formed plaza groups as well as isolated structures and clusters of structures not enclosing a central courtyard site (Robin 2012). Evidence of pole-and-thatch buildings includes casts of wooden posts and adobe walls encased in volcanic tephra at Ceren; earthen, stone, and adobe foundations; thatch-impressed clay; and dark stains of post molds where posts have decayed (Culbert and Rice 1990; Sheets et al. 2015). Evidence of perishable structures often is represented by mounded remains of successive building floors and associated artifacts, with or without stone building foundations.

A significant percentage of ancient Maya buildings are "invisible" in the modern landscape because they have stone foundations or walls buried below the ground surface or were made of pole and thatch built directly on the ground surface and have left no mounded remains (Chase 1981; Culbert and Rice 1990). Entire sites with buried middens with potsherds and other discarded artifacts, lacking any structural evidence of buildings, are "invisible" in the modern landscape, as reported in the Petén region and Arvin's Landing in southern Belize (Johnston 2004; Somers and McKillop 2005). The Paynes Creek wooden buildings—constructed directly on the ground surface without stone foundations or stones lining the exterior walls—fit the category "invisible architecture" except that they were preserved in the mangrove peat below the sea floor. Their preservation at the Paynes Creek Salt Works provides the opportunity to investigate variability in the size and placement of structures as well as their function, aided by the types and spatial distribution of associated artifacts.

Modern and Historic Wooden Architecture

Archaeologists often interpret the size, shape, and construction of ancient Maya wooden buildings and their patterning over the landscape by analogy with modern or historic Maya houses. Early Spanish explorers and missionaries, later travelers, and modern anthropologists and geographers have studied perishable structures in modern Maya villages. Diego de Landa, bishop of the Yucatán in the sixteenth century, describes houses thatched with palm, with a steep roof, a long front room for "reception and lodging of guests," and a back room, which was sometimes subdivided, for the family's sleeping quarters (Tozzer 1941).

Traditional houses in the modern Maya community of Zinacantán are constructed of wattle and daub walls with thatch roof and a packed-earth floor (Vogt 1970:30–31). Several houses are grouped together around a central patio, and the compound is fenced. The members of this domestic group include several families who farm together and carry out household tasks in the central patio. At Chan Kom, a modern Yucatecan Maya village, Robert Redfield and Alfonso Villa Rojas (1962:33–35) report different kinds of wood were used for various types of building construction. Four species were used for the corner posts and the transverse beams, and none of these species were found near Chan Kom, requiring travel 4–5 km. The walls were constructed of closely spaced poles sunk into the earth. Similar lines of closely spaced poles are reported at the modern Maya village at Cobá, Yucatán (Daltabuit et al. 1988:33, 48–49), and mod ern Kekchi villages of southern Belize (Wilk 1983, 1988).

The traditional Lacandon Maya house had a thatch roof but was open-sided, without walls (McGee 2002:43, Figure 2.2). Robert Wauchope (1938) found similarities in the pole-and-thatch construction of modern Maya houses in the Yucatán with the foundations of ancient houses he excavated at Uaxactun. Some of the earliest packed-earth floors, such as from Middle Preclassic oval structures at Cuello (Hammond 1991), have post molds formed by darkened stains from rotted house posts.

De Landa describes Maya community settlement patterns in the sixteenth century in which "in the center of the town were the temples, with beautiful plazas, and around the temples stood the houses of the chiefs and the priests, and next those of the leading men. Closest to these came the houses of those who were wealthiest and most esteemed, and at the borders of the town were the houses of the common people" (Tozzer 1941). At Zinacantán, each household's maize field surrounds the household compound, creating a dispersed community settlement pattern (Vogt 1970:31). Before modern acculturation, several Lacandon families built their houses in a group near their cornfields and separated by about a day's walk from the next settlement (McGee 2002:40).

Studies of the modern and historic Maya indicate significant variability in the use of buildings by households and the organization of economic activities, including craft production, among households. Buildings around a plaza or otherwise clustered together may include structures for different uses, including one or more residences, and other structures such as a kitchen house, storage building, or workshop.

The wooden structures formed by the patterns of wooden posts and the spatial arrangements of structures at the Paynes Creek Salt Works provide direct evidence of ancient Maya wooden architecture at the level of individual structure, clusters of buildings, and the larger community. Information about the preserved wooden structures can be used to evaluate whether ethnographic and ethnohistoric analogies used to interpret ancient Maya wooden architecture are valid: Was there was continuity in wooden architecture from ancient Maya times to the present, despite significant cultural changes over the last 1,500 years? Are the modern patterns of wooden architecture replicated in the past, or are there ancient building styles not represented by modern correlates?

The Ancient Maya Built World of Salt Production

A site was defined at the Paynes Creek Salt Works as a cluster of posts and associated artifacts visible on the sea floor. Areas between sites were devoid of visible posts and artifacts (Figure 4.3). A site may represent a household salt kitchen with larger sites including several salt kitchens and possibly residences and other special-purpose buildings. After flotation survey and flagging of posts and artifacts, the location of each post was mapped using a total station set up at one of the permanent cement datum markers created by the project. The data were downloaded and saved as UTM coordinates in Excel files and then attached to the project GIS Underwater Maya (McKillop 2012). Other information about the posts collected during underwater survey includes the type of wood, either palmetto palm or "solid wood," and the diameters of the posts. Diameters of each post were measured using a plastic covered cloth sewing tape and recorded in waterproof notebooks for subsequent transfer into the Excel files with the post coordinate information. Mapped data on posts and artifacts for each site were saved in separate files, with posts, artifacts, site boundaries, and site markers on separate sheets. The post data were copied and pasted into a single Excel file in order to visualize, compare, contrast, and manipulate the spatial data for all salt works.

Posts were discovered in areas where the floor of the lagoon is mangrove peat, sometimes overlain by silt. Only posts visible on the sea floor or discovered by feeling wood on the sea floor during survey were located, with the assumption that many posts are buried below the peat. Some underwater sites that are covered by red mangrove peat were likely flooded

Figure 4.3. Red flags mark wood posts.

quickly, submerging and killing the living mangroves. In some areas of the lagoon, such as Site 72, parts of the site are buried below living red mangroves and mangrove peat (Sills and McKillop 2010). At Orlando's, the survey team was able to find buried posts using a printed GIS map of the site to search for the existence of posts that would form a structure. In one case, discovery of a corner post meant that a triangular pattern of posts was actually a rectangular structure. In another case, posts discovered between two clusters of posts that faced each other from a distance in the shape of "open brackets" turned out to be a large rectangular structure. Many sites appeared to be mostly covered by mangrove peat, with only a few posts to indicate the presence of a structure but lacking any evident structural pattern to guide additional searching for buried posts.

Wooden Building Construction

Wood used in building construction included vertical posts, wedges for posts, and horizontal beams for a total of 4,042 mapped architectural pieces of worked wood. In addition, pieces of palm frond were found, including a fragment from the Stingray Lagoon site radiocarbon dated to the Late Classic (cal AD 690–900 and 920–950). Wood posts were mapped with a total station at 70 underwater sites, including 2,439 vertical hardwood posts, 1,482 vertical palmetto palm posts, 32 horizontal hardwood beams, 16 horizontal palmetto palm beams, and 73 wedges. Forked posts were found at Site 58 and at Site 78. The forked post from Site 58 was 2.35 m in length and 24 cm in diameter. One fork was 2.35 cm in diameter, whereas the other fork was 20 cm in diameter. The forked posts may have been supports for horizontal roof beams, for holding horizontal poles for use as fish-drying racks, or for holding salt pots over fires. Vertical wood posts were discovered at 30 sites but not mapped using a total station due to their location in a remote part of the lagoon (shallow-water Killer Bee lagoon) or in deeper water (Main Channel sites 62, 63, 64, 55, 98–110). Two sites with earth mounds are terrestrial and lack preserved wood (Watson et al. 2013). Posts were not found at seven underwater sites, although further survey might reveal posts.

Many horizontal beams were located but were not exposed, which would have led to their deterioration. The mapped horizontal beams include a measurement at each end, although some beams may have been broken, so the measurements are not of complete beams. Horizontal beams were discovered when searching for posts, often alongside a line of palmetto palm posts. Visible horizontal construction wood was mapped or its occurrence was recorded in field journals. Horizontal wood was often deeply buried beside lines of palmetto palm posts such that exposing the horizontal beams would subject them to decay. At Site 49, horizontal hardwood and palmettos were discovered deeply buried alongside vertical posts that marked the perimeter. The hardwood beam was observed to extend from posts 50 to 63. The 47 horizontal beams that were mapped include 29 made from hardwoods and 18 from palmetto palms. At Site 58 there was a 4.83 m long hardwood beam measuring 23 cm in diameter. Horizontal hardwood beams from Site 58 and from Site 67 were sharpened at both ends. Additional horizontal beams were noted during survey and mapping but were not exposed in order to preserve them.

The structures were built on mangrove peat, with other areas of unconsolidated loose silt devoid of preserved wood. At some sites a thin veneer of silt covers the peat, with the posts discolored and worm-eaten in the overlying silt but dramatically preserved in natural wood color, shape, and internal wood structure in the peat below. Ceramics are abundant at most sites, both on the sea floor and embedded in the peat below. The large

Table 4.1. Measurements of rectangular structures at the Paynes Creek Salt Works

Site, building	Estimated # pots	Width m	Length m	Floor area sq m	Room width m	Room length m	Room area sq m
7a		4.0	7.6	30			
7b		3.8	6.5	25			
7c		3.1	3.2	10			
12	90	2.2	4.1	9			
23	410	5.5	7.3	41			
32	240	3.7	6.5	24			
35	240	3.4	7.6	24			
47	360	4.4	8.1	36			
54	300	3.0	10.1	30			
58a	440	5.7	8.8	44			
58b	50	1.9	2.8	5			
60		3.2	9.0	29			
61		3.2	9.4	30			
69							
74a	350	3.5	9.9	35			
74b	160	3.3	4.9	16			
75	380	4.7	8.0	38	4.5	1.6	8.9
75					4.5	2	7.2
75					4.8	2	7.7
75					4.5	2	8.7
77a	320	5.0	6.3	32			
77b	100	2.6	3.8	10			
103	260	4.3	6.0	26			

Note: Estimated number of vessels using 10 Punta Ycacos rims/sq m excavated from Transect 1, building A at Site 74.

size of potsherds suggests there was little postdepositional site trampling (McKillop 2002).

The distribution of posts on the sea floor reveals patterns for individual structures, site patterns, and patterns for the salt works in general. There are well-defined rectangular wooden structures at some sites and rectilinear patterns of structures at many other sites. The distribution of posts at Site 35 in the East Lagoon shows a clear footprint of a rectangular structure with distinct corners (Figure 4.4a, Table 4.1). The inside dimensions of the building measure 3.4 m by 7.6 m, with an interior area of 24 sq m. The posts are clustered, with open spaces along the walls, suggesting that

Figure 4.4. Distribution and size ranges of wooden posts at Site 35 (*a*, *c*) and Site 32 (*b*, *d*).

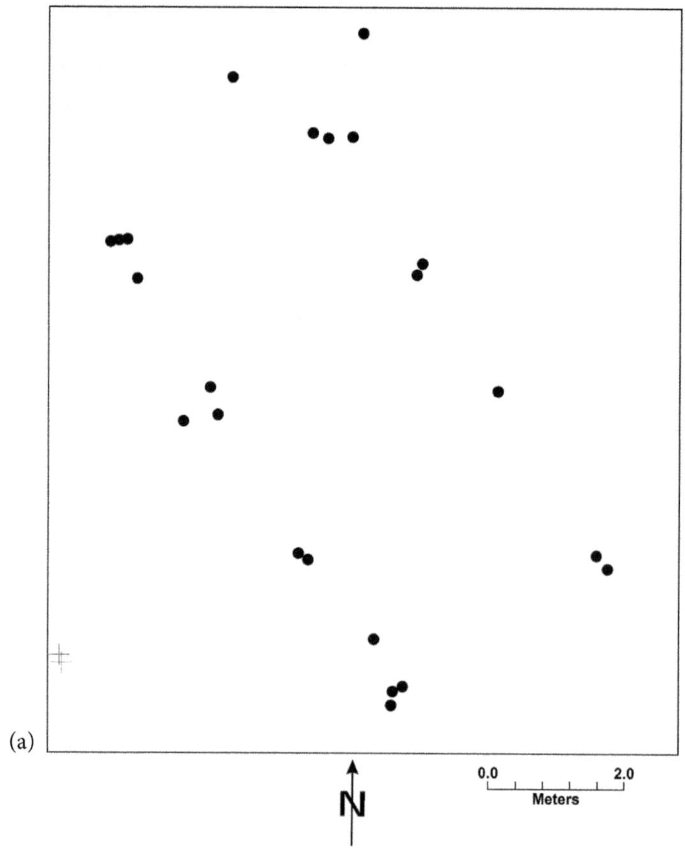

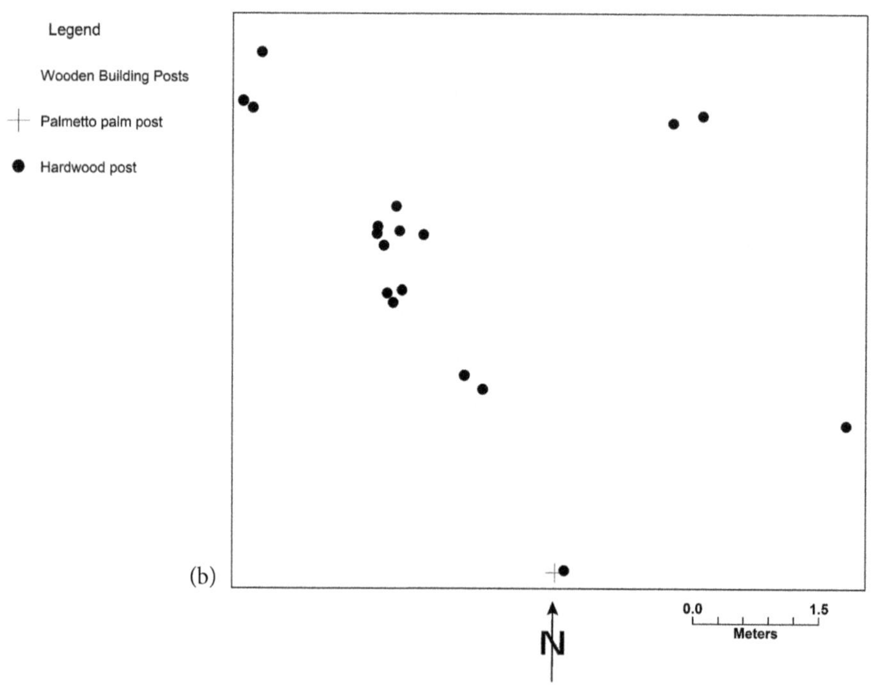

(c)
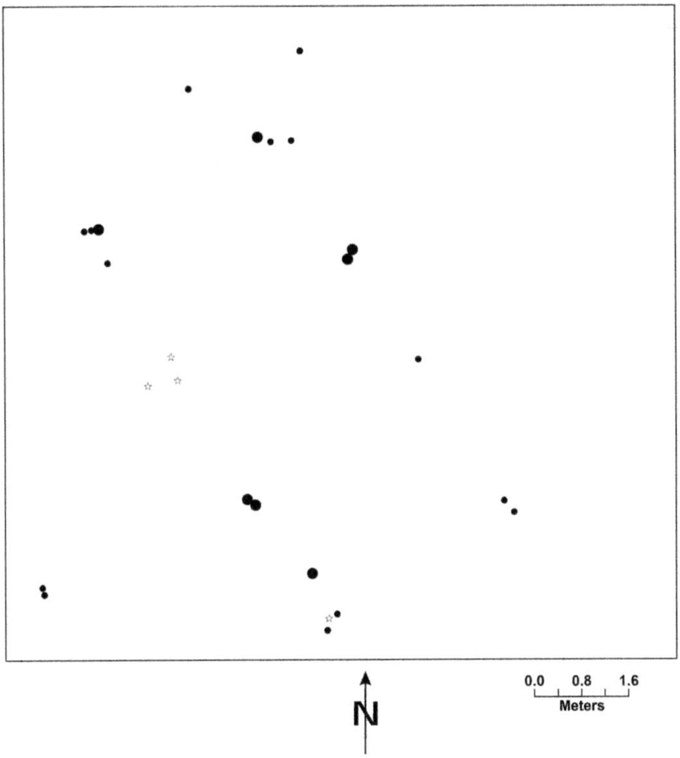
N
0.0 0.8 1.6
Meters

(d)
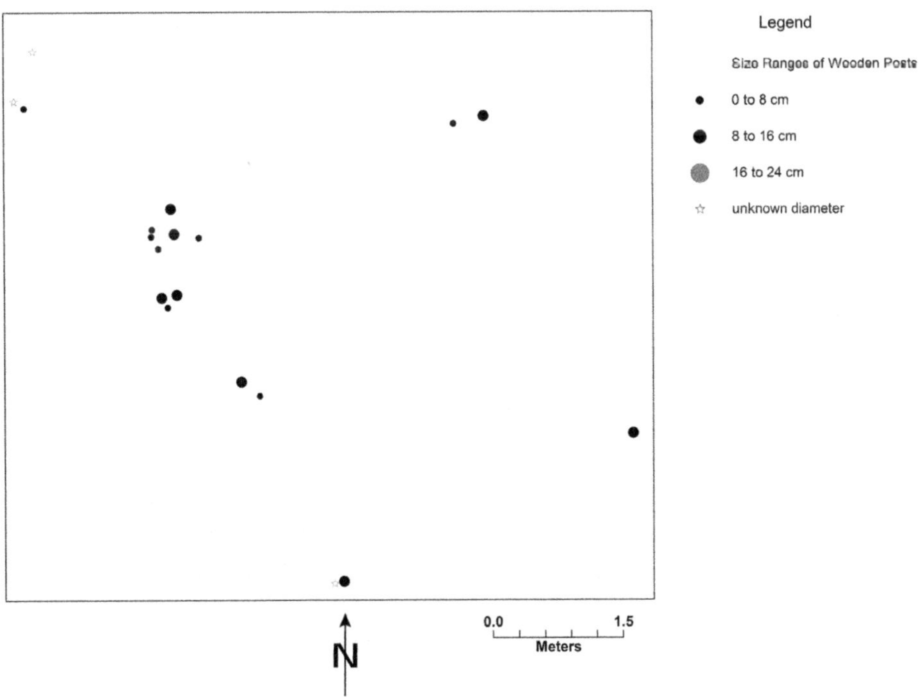
Legend
Size Ranges of Wooden Posts
0 to 8 cm
8 to 16 cm
16 to 24 cm
unknown diameter
N
0.0 1.5
Meters

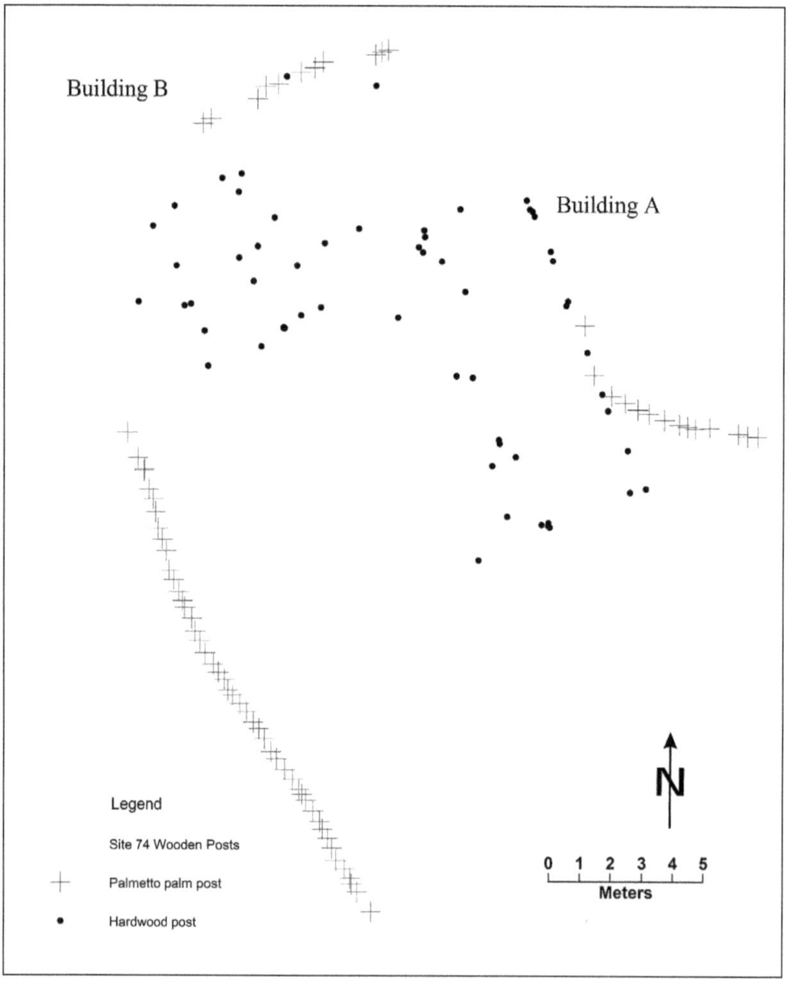

(a)

Figure 4.5. Distribution of wooden posts at Site 74 (*a*) and Site 77 (*b*), showing the rectangular structures defined by lines of dots.

either the wall posts were not dug into the ground (Wauchope 1938) or the structure had open sides, with posts at the corners and along the sides to support the roof, as described for the Lacandon Maya during modern times (McGee 2002:43, Figure 2.2) and for modern salt kitchens in Thailand (Halliwell et al. 2016). The Site 35 building is oriented in a northwest to southeast direction, an alignment that is common at the Paynes Creek Salt Works. Site 32, also in the East Lagoon, has a similar distribution of wooden posts (Figure 4.4b). The Site 32 structure measures 7.6 m by 3.4 m with an area of 24 sq m, identical to the area of the Site 35 structure. The

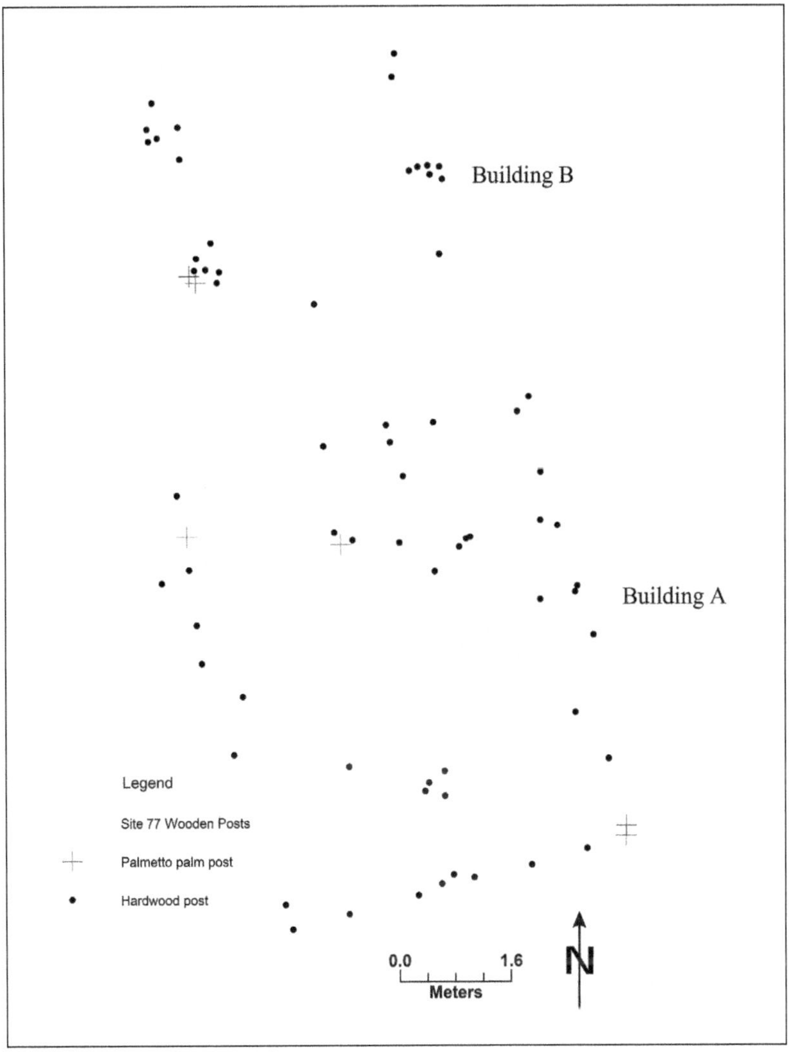

(b)

structure is aligned in a northwest to southeast direction, as at Site 35. The long sides of the buildings are oriented 32 degrees west of north.

Some sites have more than one rectangular structure, as indicated by maps of Sites 74 and 77 in the West Lagoon (Figure 4.5). They each have a large and small structure nearby. The larger structures have an area of 35 and 32 sq m, respectively, while the smaller structures have an area of 16 and 10 sq m, respectively (Table 4.1). The northern structure at Site 77 has clusters of posts similar to the pattern for Sites 32 and 35. The posts are more evenly spaced along the walls of the southern structure at Site 77

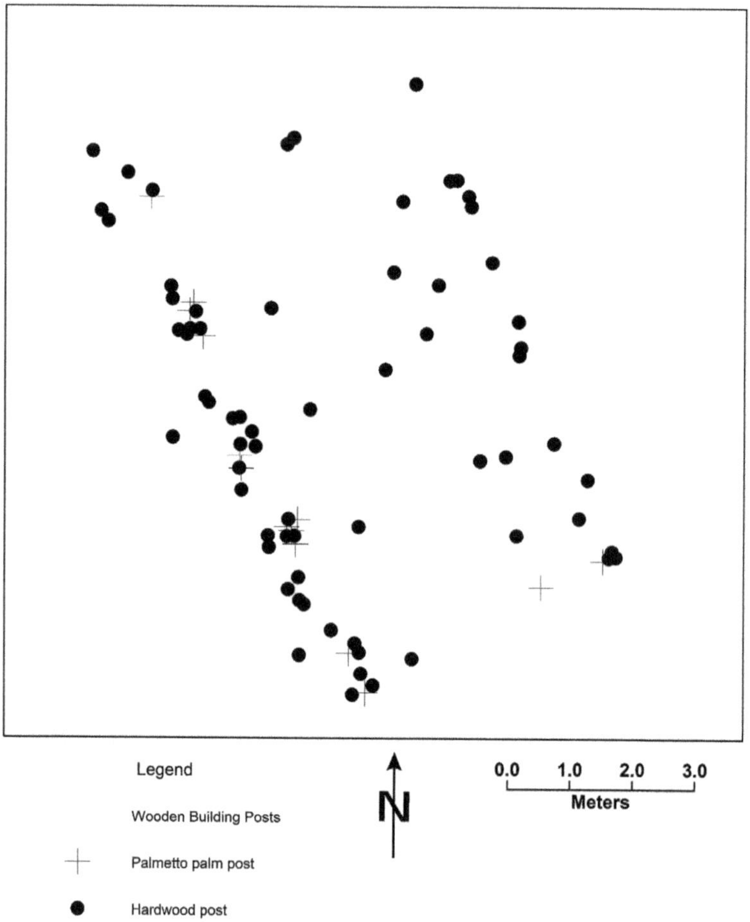

Figure 4.6. Rectangular structure defined by wooden posts at Site 75.

and the two structures at Site 74. The alignment of the structures at Sites 74 and 77 closely resembles that for Sites 32 and 35.

Some structures have interior room divisions, notably the Site 75 structure (Figure 4.6). The exterior walls are demarcated by clusters of posts. The Site 75 structure includes an interior area of 38 sq m. The building is oriented in a northwest to southeast direction, similar to buildings at Sites 32, 35, 74, and 77. There may be four rectangular rooms at site 75, with from 7.2 to 8.9 sq meters of interior space each (Table 4.1). Alternatively, there may be a roof-overhang area or long, narrow room along the eastern and western sides of the structure, with smaller rooms inside, as described by de Landa for the contact period Maya (Tozzer 1941).

Size of Wood Selected for Building Construction

Larger, load-bearing posts were expected to be at corners, as supports along the sides of buildings, and as interior supporting posts, in contrast to other wall posts, as is the case with modern Maya traditional pole-and-thatch buildings (Figure 4.2a; Wauchope 1938). The diameter of each mapped post was measured to record the size of posts and to discover patterns within buildings and in different areas of the lagoon system. The posts were measured in situ by first removing peat by hand from around the circumference to measure firm wood, in contrast to the worm-eaten wood that was exposed above the sea floor. The circumference was measured instead of the diameter when the top of a post was uneven, making a diameter measurement awkward. Some posts could not be measured for a variety of reasons. Care was taken to ensure that measuring diameters underwater was providing accurate information. Average diameter of the sample of hardwood posts is 8.4 cm, which is larger than the average diameter of 6.0 cm for palmetto palm posts.

I grouped post diameters by size ranges in the Underwater Maya GIS to discover patterns of construction. Grouping post diameters by the average size ranges reported by Wauchope (1938) in his study *Modern Maya Houses* revealed similar patterns between modern and ancient construction techniques: Larger, "load-bearing" posts are located in the corners and exterior walls, with smaller posts located along the exterior walls and inside structures. The same pattern is evident in the Robert C. West photo (Figure 4.2a). Plotting posts by size intervals shows that most posts are 8–16 cm in diameter, with some less than 8 cm in diameter. Large posts were rarely used, indicating young trees were selected. Post exteriors were not cut or otherwise shaped, and bark often was present, indicating posts were likely cut and used in construction while the wood was still green so that the bark had not fallen off the tree.[2] There are few posts 16–24 cm in diameter. Most posts are 0–8 or 8–16 cm in diameter. The pattern is evident for rectangular buildings in different areas of the lagoon system, including the West Lagoon, East Lagoon, Stingray Lagoon, the Main Channel, and West Point (Figures 4.7–4.13).

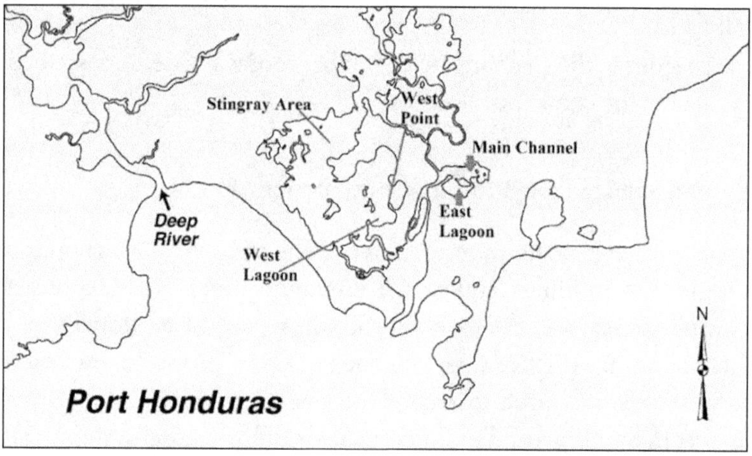

Figure 4.7. Map showing the five areas of the Paynes Creek Salt Works.

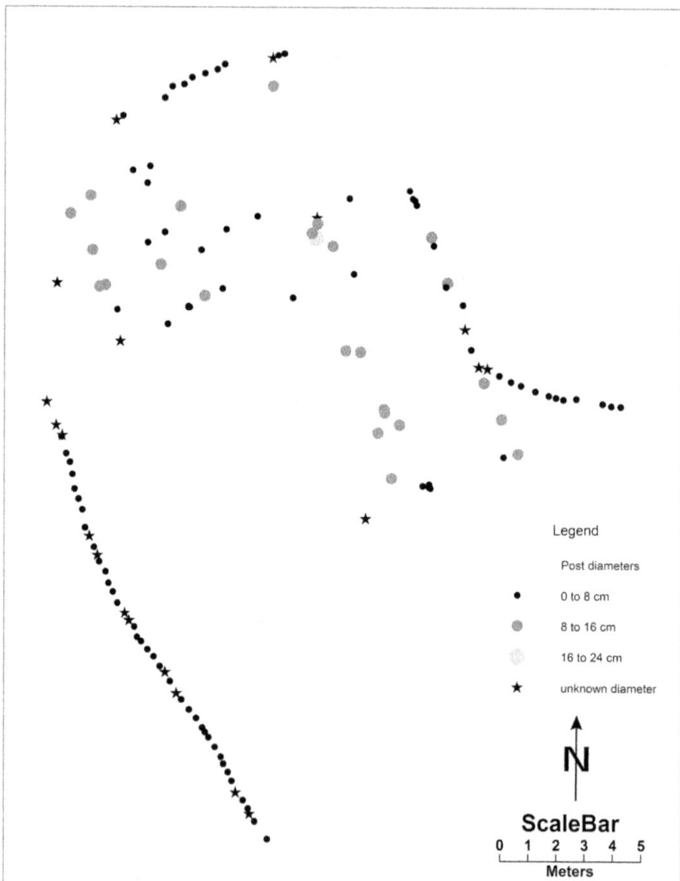

(a)

Figure 4.8. Size ranges of wooden posts at Sites 74 (*a*), 77 (*b*), and 75 (*c*) in the West Lagoon.

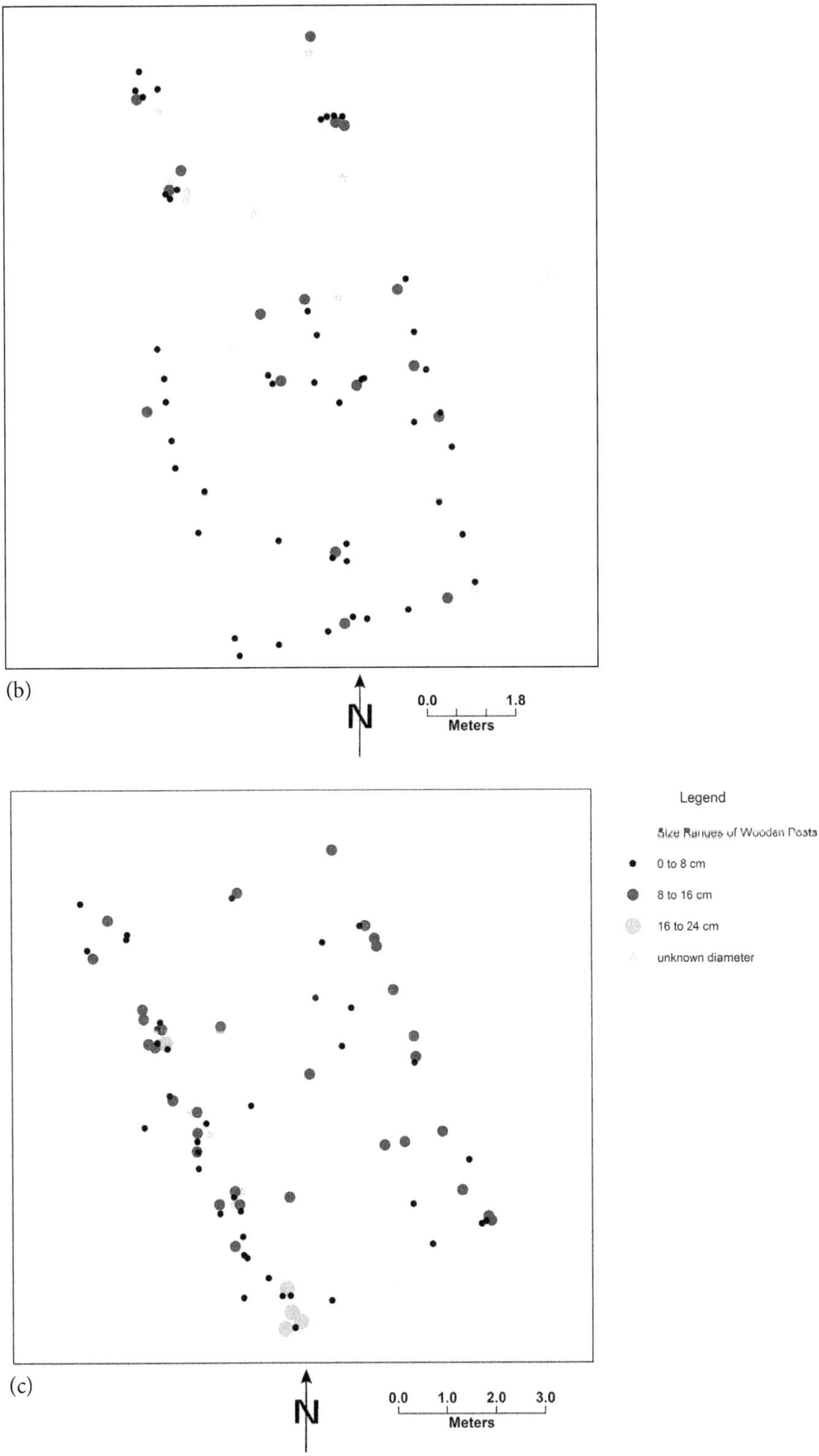

0.0
1.8
Meters
N
0.0
1.0
2.0
3.0
Meters
N
Legend
Size Ranges of Wooden Posts
0 to 8 cm
8 to 16 cm
16 to 24 cm
unknown diameter

(b)

(c)

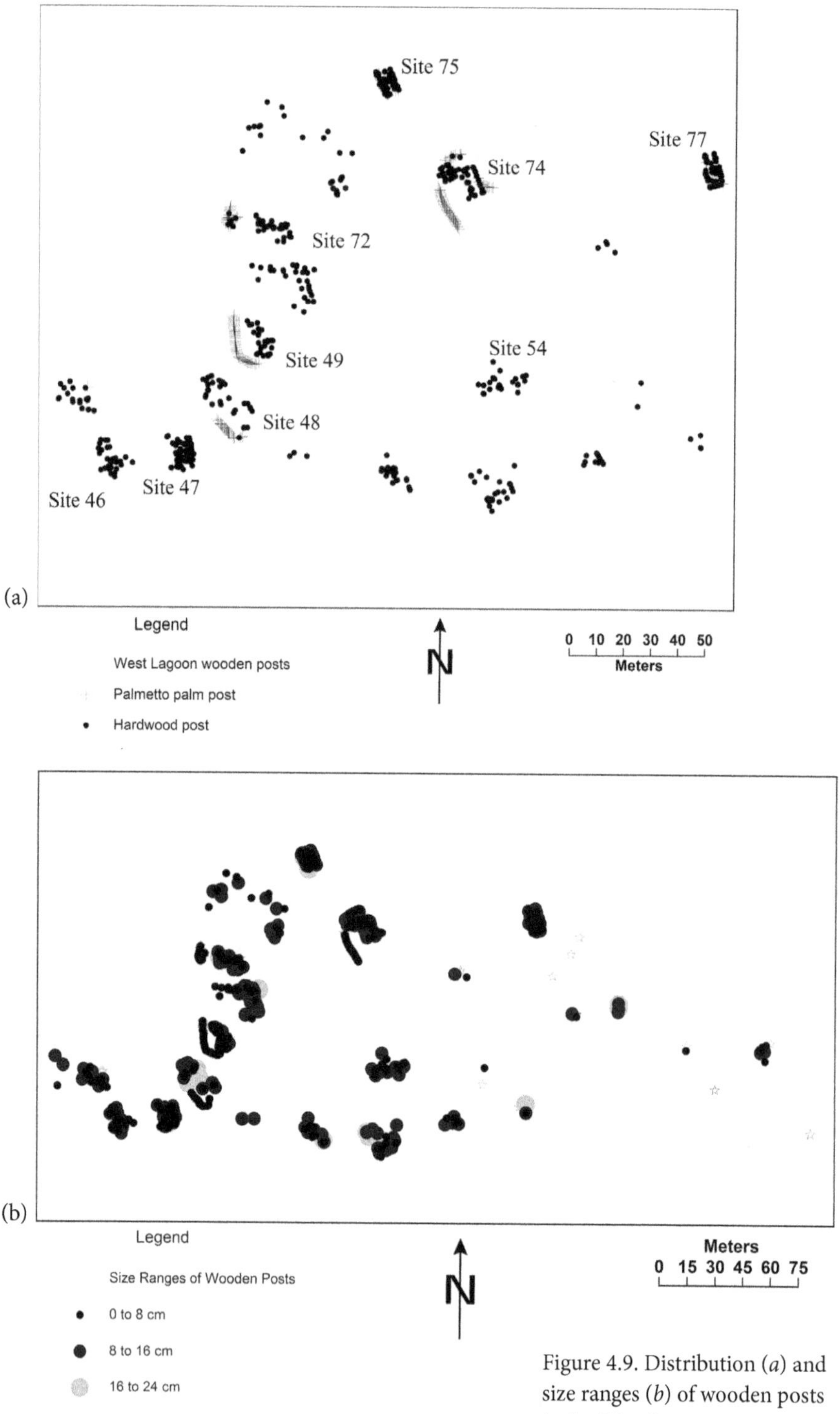

Figure 4.9. Distribution (*a*) and size ranges (*b*) of wooden posts in the West Lagoon.

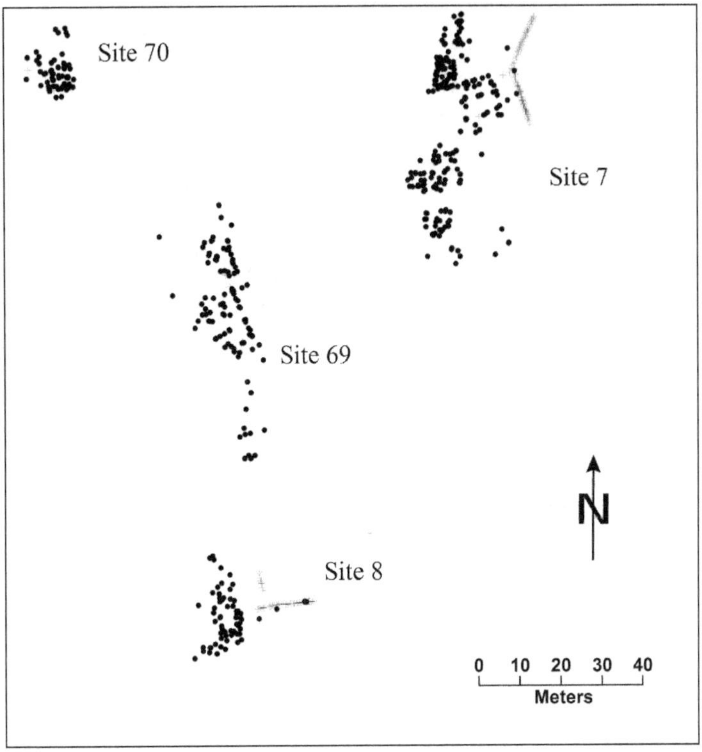

Figure 4.10. Distribution of wooden posts in the West Point area.

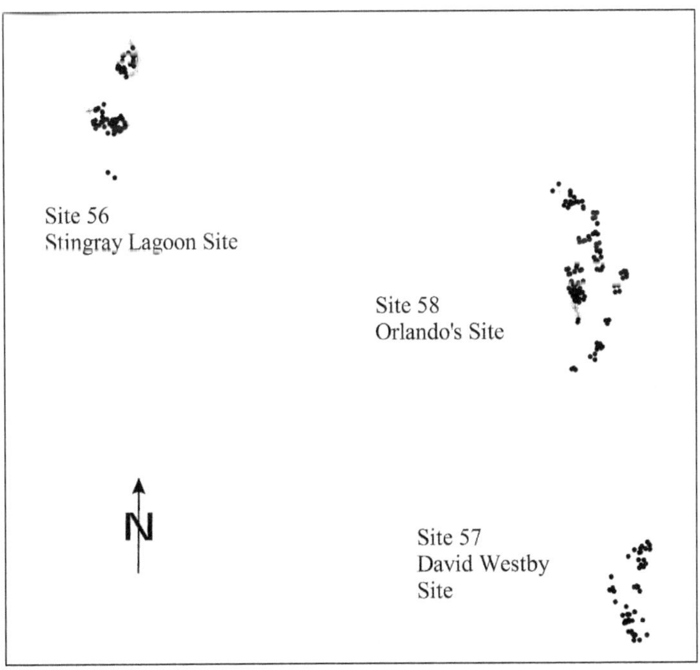

Figure 4.11. Distribution of wooden posts in the Stingray Lagoon area.

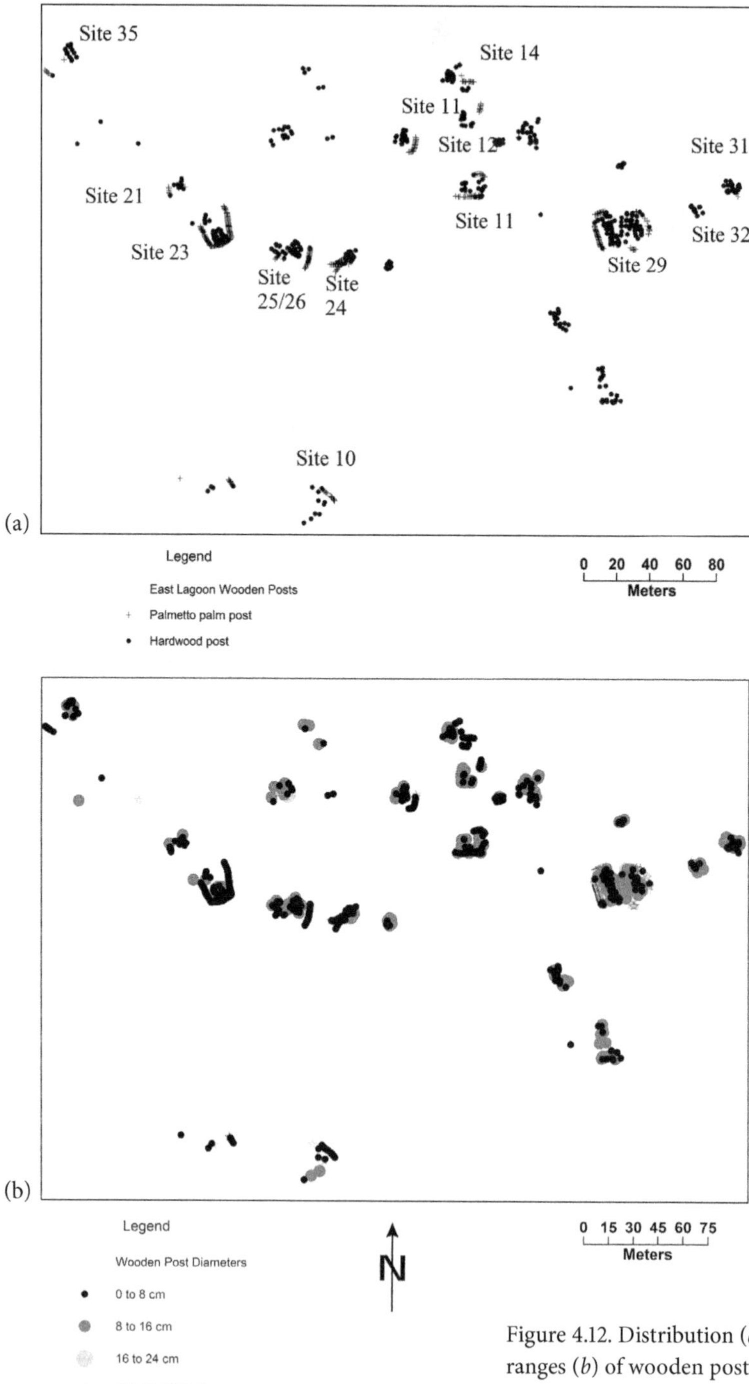

Figure 4.12. Distribution (*a*) and size ranges (*b*) of wooden posts in the East Lagoon.

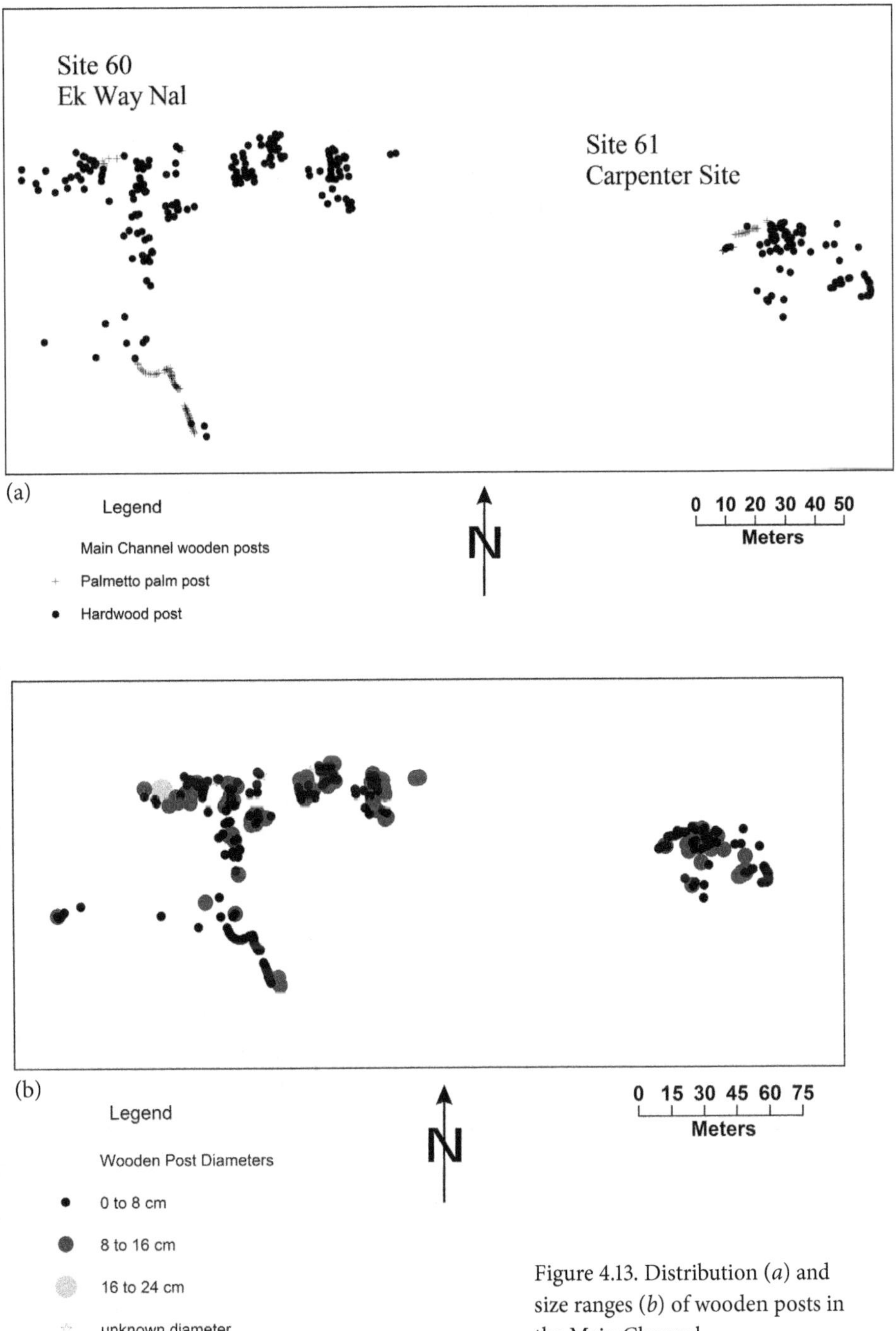

Figure 4.13. Distribution (*a*) and size ranges (*b*) of wooden posts in the Main Channel.

Post Size Selection in the West Lagoon

The most clearly delineated wooden building outlines are in the West Lagoon. Most building posts for structure 74a are 8–16 cm in diameter, with some less than 8 cm (Figure 4.8a, Table 4.1). The smaller structure, 74b (4.9 × 3.3m), has the same repertoire of posts on corners and along sides. In addition, there are posts in the interior that may have supported the roof or divided the space into two rooms as well as posts to support a roof overhang on the east side. The largest posts are in the center of the building (12 cm diameter) and the center of the east wall (13 cm diameter).

Site 77 has two wooden buildings, including structure 77a on the south (6.3 × 5 m) and a smaller structure, 77b (3.8 × 2.6 m), on the north (Figure 4.8b). Posts in the interior of structure 77a may be small storage areas or may represent furniture supports, as with the elevated hearth for evaporating brine inside buildings at Sacapulas (Reina and Monaghan 1981). A cluster of posts may partition the south end into an area 1.3 × 5 m or even two smaller areas 2.5 × 1.3 m each. Similarly, interior posts may create partitions on the north end of structure 77a, using the north 2 m for two to three areas. More striking are the larger diameter interior posts that are aligned with posts on the north and south walls. These larger posts may have divided the building into two rooms or may have been support posts inside the building. Structure 77b has clusters of posts at each of the four corners, with larger posts along with smaller ones. The lack of wall posts suggests building 77b was an open-sided structure.

Structure 75 is a four-room building with larger corner posts and interior posts dividing the structure into rooms (Figure 4.8c). The plethora of smaller posts along the western side may be posts for a roof overhang or the posts created a long, narrow room. The overall building footprint of posts indicates a rectangular structure of 8–9 × 4.5–5.5 m. The rooms each measure about 8 × 2 m. Some of the smaller diameter posts along the exterior of structures, such as the western side of the Site 75 structure, may be roof poles that supported the overhang of the thatched roof that extended beyond the exterior walls. An overhang is especially important to keep rain out if the sides of the structure are open. There are more posts along the western side of the Site 75 structure, which is perplexing. They include palmetto palm posts, which may have been used as supports for roof overhang. On the west side of the structure, palmetto palm posts are associated with clusters of solid-wood posts. On the south side is a single palmetto palm post along the wall as well as one at each corner associated with solid-wood posts. Perhaps there was more need for support of a roof overhang at the corners of a structure than midway along a wall.

There are several additional rectilinear buildings in the West Lagoon, but none is as clearly defined as those at Sites 74, 75, and 77 (Figure 4.9, Table 4.1). Structure 47 has a mass of posts that form a rectangular shape, suggesting building renovation or wooden supports for furniture inside the building. Sequential construction episodes or posts that were supports for tables or drying racks for fish inside the building may explain the distribution of posts. Large areas of Site 72 are covered by mangrove peat or living mangroves, obscuring the building outlines, although some rectilinear patterns are evident.

Post Size Selection Elsewhere at the Salt Works

Northeast of the sites in the West Lagoon are sites off a modern point of land referred to as West Point (Figure 4.10). Site 7 has over 400 wooden posts that are clustered in several areas indicating there are multiple buildings, some with clear rectangular outlines formed by posts. There are as many as 10 structures and three plazas (Table 4.1; McKillop 2018:Figure 4). Two lines of palmetto palm posts form a long line along the eastern boundary of the site. The palmetto palm posts served as a retaining wall. East of the wall was devoid of artifacts, and the sediment was silty in contrast to the firm mangrove peat that forms the matrix of the site. Site 69 has clusters of posts that loosely mark the location of two buildings, although their size and number of rooms are poorly defined by the post distributions. One building either has interior posts or room partitions.

There are three sites in the Stingray Lagoon area (Figure 4.11). Site 56 (Stingray Lagoon site) on the west has three areas of poorly defined building post footprints in the southern area of the site. There is a line of palmetto palm posts to the east of two buildings and to the west of the third. The posts are similar in size. There is a poorly defined area of hardwood posts in the north area of the site that may form a rectangular building west of a line of palmetto palm posts. Sites 57 (David Westby site) and 58 (Orlando's site) are 300 m to the east near an island. Site 57 (on the south) has a line of posts that may be part of a structure. Site 58 includes several buildings as well as other structures that may include brine-enrichment supports, an idea derived from the small clusters of posts.

In the East Lagoon, there are clearly defined rectangular buildings at Sites 35, 32, 23, and 25/26 as well as less clearly defined buildings at Sites 29, 24, 13, and 11 (Figure 4.12). Corner posts are in the 8–16 cm diameter range. At Site 23 there are larger posts that demarcate a rectangular building footprint. There may be a building with a side room or overhang to the west and to the east. Smaller posts to the north and south may also be roof-overhang posts. At Site 25/26 one building has larger corner posts than posts along the north and south walls. The two sites were combined into a single site after survey indicated artifacts and wooden posts continued between the sites. There are several buildings at Site 29, but their sizes are not clear. Looking at the distribution of larger posts, there may be one structure parallel to the line of palmetto palm posts on the west side of the site. There are certainly other buildings at the site. Site 24, Chan b'i, has at least one building with room divisions (Sills and McKillop 2013, 2018a; Sills et al. 2016).

Of the sites in the Main Channel where the water depth often exceeded our ability to map with a total station and required diving with air hoses, there are two mapped salt works as well as sites in deeper water that also have shell middens (Figure 4.13; McKillop 2017a). At Site 60 (Ek Way Nal), there are several areas of the site with posts 8–16 cm diameter that demarcate rectangular building footprints (Figure 4.13). There are two larger posts, each 17 cm diameter, that are outside one building and in the center of another. The larger 8–16 cm diameter posts define long, narrow buildings. However, none is as clear as the buildings at Sites 74, 75, and 77 in the West Lagoon. Site 61, the Carpenter site, has a T-shaped structure, 61a, that may include side rooms as at Site 75. The corner posts and major posts spaced along the east and west walls are 9–11 cm in diameter, with a large 18 cm diameter post at the southwest corner and midway along the east wall.

Salt Kitchens at the Paynes Creek Salt Works

The Paynes Creek Salt Works comprise a large salt-production complex with buildings spread out over about 10 sq km with most sites concentrated within a 5 sq km area. During survey, sites were identified by clusters of artifacts and posts on the sea floor that were separated by open space from other clusters. Some buildings were identified during survey from the patterns of flags we placed to mark post locations. For many sites, the distribution of posts was not clear until the total station map data were downloaded and added to the project GIS map. The maps provided an opportunity to see if additional posts could be located to complete building footprints. We printed site maps and put them in ziplock bags, returned to sites, and searched for missing posts.

After several seasons of flotation survey and total station mapping, additional patterns were evident: some sites had more than one building, and most buildings were oriented 37 degrees west of north along the long side. There are two different settlement patterns evident from post distributions: there are a few large sites with multiple buildings, such as Sites 7, 60, and 72. By way of contrast, other buildings were in lines along ancient shorelines. This pattern resembles salt works along the Gulf of Fonseca on the Pacific coast of Costa Rica (Bishop 1977:Figure 39) and coastal salt works in Shandong Province in China (Flad 2013). The rectangular structures vary in size (Table 4.1), consistent with research elsewhere in the Maya area, indicating that the size of structures varies tremendously according to family size or use as kitchens or workshops. Some sites have lines of posts that clearly formed a structure, but the outline shape is unclear. Roof posts outside the wall posts may obscure the outline of the structure.

Lines of Palmetto Palm Posts

The palmetto palm (*Acoelorraphe wrightii*) posts are easily identifiable from other "solid-wood" posts. During survey and mapping, posts were identified as solid-wood or palmetto palm posts. There are lines of palmetto palm posts at 17 sites (Figure 4.14, Table 4.2). Some lines are up to 10 m away from the wooden architecture, creating an open space. Chac Sak Ha Nal (Site 23) has a line of palmetto palm posts forming a "U" shape around a structure formed by wooden posts. Some of the lines of palmetto palm posts abut the wooden architecture.

A building at Site 103 has two walls formed by lines of closely spaced palmetto palm posts. The corner posts are hardwood. As with Site 75, there is a long, narrow space along the east side that may be a corridor or roof overhang. There is one large interior post measuring 17 cm diameter. Another building at Site 103 abuts the north line of palmetto palm posts

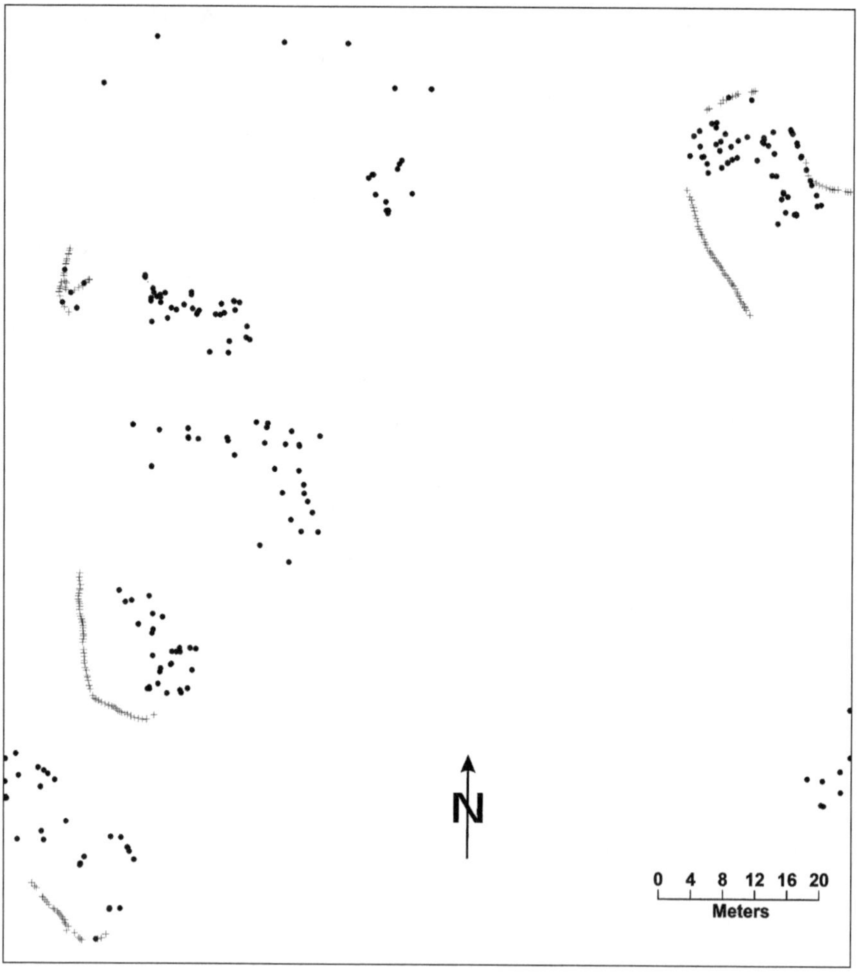

Figure 4.14. Lines of palmetto palm posts at selected West Lagoon sites.

and is defined by six hardwood posts. The building may be much larger, depending on how the lines are drawn to form the building. Each building has an open area defined by palmetto palm posts.

Are there patterns in the lines of palmetto palm posts that indicate their use, perhaps as land-retaining walls, household/workshop boundary fences, or walls around salt pans? The palmetto palm posts were difficult to find since they often form lines up to 10 m away from the architecture

Table 4.2. Measurements of the lines of palmetto palm posts

Site	Length m	Defines yard	Edge of site	Abuts building
74	17.6	yes	yes	yes
	6.5			
	5.2			
7	14.9	yes	yes	
	13.9			
58	12.9	yes	yes	yes
60	23.1		yes	
	9.8			
61	4.2		yes	
	8.3			
	10.3			
103	16.3		yes	
	14.1			
	9.0			
	19.6			
15	10.1	yes	yes	
25	13	yes	yes	yes
26	3.2			
72	8.2		yes	
	4.9			
8	4.6	yes	yes	yes
	14.3			
13	5	yes	yes	yes
14	7			
49	23.6	yes	yes	yes
48	12.7	yes	yes	yes
23	46	yes	yes	yes
	29			
	6.5			
	15.4			
21	5	yes	yes	yes
35	3.2			
11	7.5			
	15.9			
	3.1			
5	4.9			
10	11.1			
67	10.2			
	13			
	20.9			
	12.4			
	4.6			
	9.0			
56	25.9			
	12.5			
	3.3			

and only identifiable by the bark—a hard shell with diagnostic fibers that is impervious to salt water. Where lines of posts were discovered, the record is good since a lot of effort was made to find all the palmetto posts. Only at Site 8 did we abandon the search for the end of a line of palmetto palm posts during survey. At Site 8, one of the lines of palmetto palm posts continued toward the modern shoreline but was progressively more deeply buried in the peat (Figure 4.10). To continue uncovering the deeply buried posts would then expose them to deterioration.

If the lines of palmetto palm posts were used for land retention at the water's edge, there should be similar lines at adjacent structures and sites. The lines of posts should be palmettos in preference to other tree species that are water resistant. If the lines of posts served to define space or household/workshop boundaries, they should enclose one of more structures and include an open space. Furthermore, the lines of posts would not necessarily be arranged in the same linear fashion at adjacent sites. If the lines of posts were salt pans, as on the north coast of the Yucatán (Andrews 1983:Figures 2.3–2.4), the posts should enclose a space.

Available evidence indicates that the lines of palmetto palm posts were used to retain land and that the space defined an outside work area beside wooden buildings. Most of the sites are defined by wooden architecture and associated artifacts on the sea floor. In some cases, there is a line of palmetto palm posts at some distance from the structure(s) but at the edge of the artifact distribution. As such, the lines of palmetto palm posts define an open space outside the building. Structures forming a linear pattern have lines of palmetto palm posts on the same side and similarly oriented, suggesting placement for land retention. The lines of palmetto palm posts vary in length, to a maximum recorded length of 46 m at Site 23 (Table 4.2). The posts are generally closely spaced. Short cores at Sites 24 and 35, taken to see if marine sediment occurred on one side of the lines of palmetto palm posts and mangrove peat on the other side, were inconclusive largely due to sediment accumulation since the sites were submerged.

The visible lines of palmetto palm posts do not form a polygon, as would be expected if they were used to define the perimeter of an evaporation pan. Individual palmetto palm posts were incorporated into buildings at some sites, but this use was infrequent. At Site 103 the lines of palmetto palm posts form parts of the walls of the buildings. Perhaps they were preserved wall posts that did not preserve at other sites since they were resting on the ground surface instead of being dug into the ground. Certainly at least some of the palmetto palms posts were deeply buried in the ground, such as the perimeter wall at site 74. Most of the lines of palmetto palm posts form curved lines, although at Site 103 there are four straight lines of posts, each including a corner. Two of the lines have right-angle corners. Evidently the waterproof nature of the palmetto palm bark was regarded as making the posts useful for land retention at the water's edge.

Community Settlement Patterns

Were some or all of the Paynes Creek buildings salt kitchens for evaporating brine to make salt as documented elsewhere in Mesoamerica in historic and modern times? Were there houses, kitchens, and nondomestic civic structures? Was there a "cottage industry" of salt production associated with Maya households, as described for chert tool production at Colha in northern Belize (Shafer and Hester 1983)? Previous excavations at four salt workshops revealed they lacked the range of vessel forms used as containers at Maya communities, they lacked burials or remains of houses, and they had a restricted array of pottery types—only five (McKillop 2002).

Few of the Paynes Creek sites are organized as *plazuela* groups—buildings arranged around plazas that were the building blocks of ancient Maya settlements, from public stone architecture in the center of cities down to modest domestic household groups. Most structures form lines instead of grouping to form a central plaza. In both the East and West Lagoons, there are several lines of structures, with open space in between (Figures 4.15–4.16).

The linear settlement pattern may reflect structures located along shorelines that were flooded, so new structures were built farther away from the water on higher ground. If salt works were relocated due to rising seas, then there should be differences in dates of building construction and of associated artifacts between lines of structures. This hypothesis will be investigated further in the discussion of the age of the structures, patterns of tree species used in construction, and associated artifacts.

Structures closest to the water may have been salt works, whereas structures farther away may have been residences. Elsewhere, salt workers do not always reside at the salt works. Parsons (2001) describes residences in central Mexico at some distance from the salt works. The location of

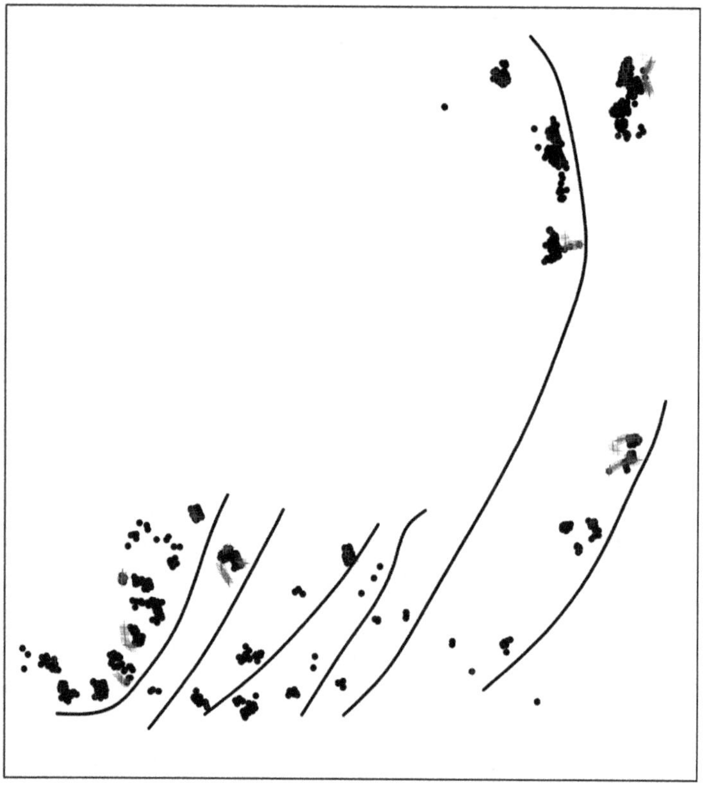

Figure 4.15. Linear arrangement of wooden buildings in the West Lagoon.

the residences for the salt workers at the Zhongba site in the Three Gorges River area of China was unknown, perhaps flooded after the excavations of the salt works (Flad 2007). Salt workers who extracted salty soil from the salt springs at Sacapulas in the highlands of Guatemala had salt sheds near the salt spring with separate residences farther back (Reina and Monaghan 1981). Differences in the artifact assemblages will be examined in chapters 6 and 7. Residences would be expected to have a wider variety of vessel forms and various domestic objects, in contrast to salt works

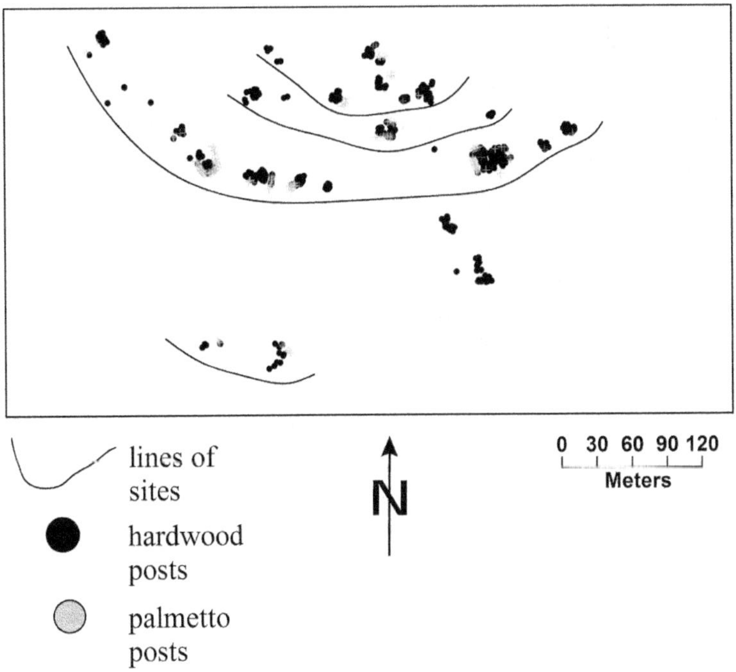

Figure 4.16. Linear arrangement of wooden buildings in the East Lagoon.

with a focused activity on evaporating salty water by heating brine in pots over fires. Artifacts recovered from different structures provide insights into the use of buildings as residences or salt works.

The lines of structures may reflect structures located on all available land, where most of the area was water. Although now underwater, the sites are located on firm mangrove sediment that likely was dry land at the time when the salt works were in operation. Between the sites there is loose silt that may represent areas formerly covered by salt water. Salt water was used for obtaining brine to boil at the salt works. Water access was also needed for transporting the salt and the salt workers. If the salt works were used at the same time, differences in the dates of structures would not correspond to the lines of buildings. This hypothesis will be investigated in chapter 6.

Summary

The Paynes Creek Salt Works include rectangular pole-and-thatch buildings that resemble modern and historic Maya construction. Different kinds of wood were used for different parts of buildings, some of which was brought from 3–4 km away, a practice also noted for the modern Maya community of Chan Kom (Redfield and Villa Rojas 1962:33–35). The structures have packed earthen floors with no evidence of plastering. Patio groups formed by buildings creating an open, central space are rare at the salt works, except for Sites 7 and 60, which are larger and have multiple buildings. Archaeologically isolated structures are common in addition to plaza groups. Site 75 has a long anterior room and several rooms behind, which resembles de Landa's descriptions of public and private space, respectively (Tozzer 1941). The linear arrangement of most buildings along ancient shorelines underscores the placement of sites for salt production. Larger sites with plaza groups may have had additional activities in addition to salt production, including residential use and multicrafting.

5

Wood Selection for the Salt Kitchens

Although sea-level rise has submerged the Paynes Creek Salt Works, the tree species available to the ancient Maya from the deciduous rainforest, pine savanna, coconut woodland, and mangrove ecosystem are those found in the area today (Wright et al. 1959). Sediment coring indicates that the ancient landscape in the immediate area of the salt works was a mangrove ecosystem dominated by the same species of red (*Rhizophora mangle*), black (*Avicennia germinans*), and white (*Laguncularia racemosa*) mangroves that characterize the modern landscape (Figure 5.1; McKee 2011; McKillop et al. 2010a, 2010b). The cores consisted of red mangrove peat to varying depths from 4 to 7 m below the sea floor. Since red mangrove peat is created when mangroves grow taller to keep their leaves above water as a result of sea-level rise, the peat is a proxy for sea-level rise in antiquity. Rapid sea-level rise in the Postclassic (AD 900–1500) flooded the red mangroves and submerged the salt works.

The mangrove ecosystem that dominated the ancient landscape included red mangroves that can grow in salt water, black mangroves that grow on tidal flats, which can be highly saline, and white mangrove that grows farther back from the sea (Figure 5.1a). Black mangroves tolerate higher salinity than red mangroves. Other trees in the mangrove ecosystem include buttonwood (*Conocarpus erectus*). The modern landscape of Punta Ycacos Lagoon is bounded by areas of higher ground that support tree species more common in deciduous or evergreen rainforests, such as sapodilla (*Manilkara zapota*). Abutting the mangrove ecosystem on the seaward side is coconut woodland that hosts palmetto palms, nance/craboo (*Byrsonima crassifolia*), cashew (*Anacardium occidentale*), icaco plum / coco plum (*Chrysobalanus icaco*), myrtle (*Myrica cerifera*), holly (*Ilex* sp.), and calabash (*Crescentia cujete*), in addition to non-native introduced species such as coconut palms (*Cocos nucifera*), breadfruit

(a) 

Figure 5.1. Mangrove ecosystem from the Main Channel toward the East Lagoon (*a*), with profile showing organic matter below the seafloor (*b*). By Mary Lee Eggart.

(*Artocarpus altilis*), bananas (*Musa* sp.), sugarcane (*Saccharuum officinarum*), and citrus trees (*Citrus* sp.). The northern and eastern shores of Punta Ycacos Lagoon abut the pine savanna, composed of open grassland with scattered clumps of trees (locally known as broken pine ridge) including palmetto palms, pine (*Pinus caribaea var. hondurensis*), sandpaper trees (*Curatella americana*), calabash, oak (*Quercus oleoides*), coco plum, madre de cacao (*Gliricidia sepium*), and black poisonwood / chechem (*Metopium brownie*). To the west of Punta Ycacos Lagoon, the change in geology to sandstone, mudstone, and siltstone marks a change in vegetation patterns, with deciduous forest species and cohune palms (*Orbigyna cohune*). The deciduous rainforest includes rosewood (*Dalbergia stevensonii*), Santa Maria (*Calophyllum antillanum*), mahogany (*Swietenia macrophylla*), sapodilla, Nargusta (*Terminalia amazonia*), and black cabbage bark (*Lonchocarpus castelloi*), among others (Wright et al. 1959).

Pollen from the cores indicates that there were various tree species typical of the modern landscape (Jones 2013; McKillop and Jones 2018; Wright et al. 1959). Pine was not used for architectural wood but is recorded in

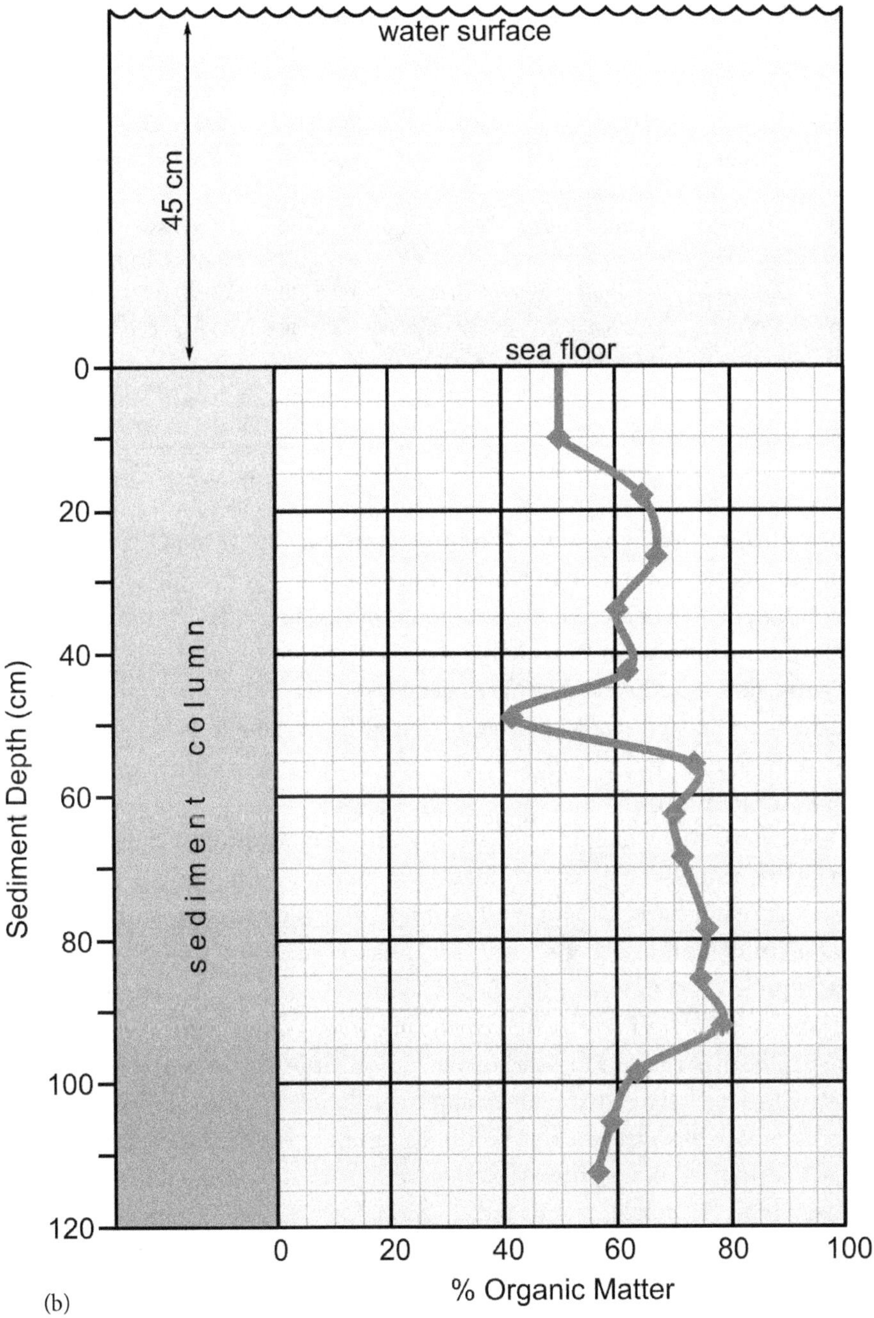

water surface
45 cm
sea floor
sediment column
Sediment Depth (cm)
0
20
40
60
80
100
120
0
20
40
60
80
100
% Organic Matter

(b)

Table 5.1. Plant remains at the Paynes Creek Salt Works

Site	Item #	Species	Common name	years BP	Calibrated, 2 Sigma C14 date (Beta Analytic Inc)
56	Mammee	*Calocarpum mammosum*	Mamey	1460±40	Cal AD 550–660
56	Palm stem	Unknown palm	Palm frond	1240±40	Cal AD 690–900 and cal AD 920–950
67	Calabash	*Crescentia cujete*	Calabash	1360±40	Cal AD 620–690
56	Coyol	*Acrocomia mexicana*	Coyol, supa	1220±40	Cal AD 680–890
56	Cohune	*Orbignya cohune*	Cohune	1280±40	Cal AD 660–810
56	Calabash	*Crescentia cujete*	Calabash	1230±40	Cal AD 680–890

the pollen record. Pine is a common tree on the pine savanna, abutting the lagoon system on the northeast. Pine was used by the ancient Maya for torches and was commonly brought or traded to ancient communities if it was not locally available (Lentz et al. 2005; Morehart et al. 2005). As a fast-burning wood, pine may have been used to start fires and would have turned to ash, so it would not be recovered from the salt works if it had been used. Pine was not identified in wood charcoal sample (Robinson and McKillop 2014).

Native palms, which are not wind-pollinated, also were common in the ancient landscape, as indicated by abundant palmetto palm posts used as retaining walls and by the ubiquity of the endocarps of native palm fruits found during survey (McKillop et al. 2010a, 2010b). Native palm pollen was not identified in the sediment cores. Both cohune (*Orbigyna cohune*) and coyol (*Acrocomia mexicana*) endocarps were common. One of each was radiocarbon dated from Stingray Lagoon site to the Late Classic, along with a mamey apple seed (*Calocarpum mammosum*), and calabash bowl fragments (Table 5.1). Roofing thatch needed to be cut and transported dry, so it was likely moved from stands of palm trees by boat. Stems from palm fronds radiocarbon dated to the Terminal Classic were found at Site 56, but the palm species is unknown. Cohune palms are common in the deciduous rainforest along the shores of Muschamp Creek and west of Deep River. The native palm endocarps as well as those from poknoboy palms (*Bactris major*) and seeds from mamey and nance/craboo were commonly found at Classic period sites on offshore islands in Port Honduras, including Wild Cane Cay, Tiger Mound, Pelican Cay, and Frenchman's Cay (McKillop 1994, 1996b, 2005b).

Methods for Identifying Tree Species from Waterlogged Wood

Few posts were excavated since they would decay once removed from the mangrove peat. After the first post was excavated in 2004 to verify that it was a post, some additional posts were excavated to verify that they were posts or because they were accidentally removed while clearing peat from around the post for wood sampling.

Wood samples were cut from building posts to identify the tree species used in construction, patterns of tree species selection for different parts of buildings, resource selection from different nearby and distant ecosystems, and changes over time in tree species or size that might indicate overuse of the rainforest. Excluding palmetto palms, which are easily identifiable, an attempt was made to sample all other posts in the 2005–2007 field seasons, with some additional posts sampled in later field seasons. The number of posts per site sampled varied from 1 post to 122 posts from a total of 56 sites. The number of posts at sites varied, so even when all posts were sampled, there was variation in the number of samples per site. Some posts were not sampled because the wood was too hard to cut, the label on the flag identifying the post number had washed off, or the site was too deep to sample without dive equipment. Wood samples were not routinely taken in 2008 and subsequent years because of the backlog of samples awaiting species identification.

Wood samples were cut from the top of posts using 18" stainless steel kitchen knives, following initial sampling with machetes. To obtain a structurally intact piece of wood (devoid of worm holes), the mangrove peat was cleared by hand from around the post 5–10 cm below the sea floor. After cutting, each post sample was placed in fresh water in a ziplock plastic bag with a label written on the bag in black indelible marker indicating the site, post number, date, and excavator. Since the wooden posts were salt-waterlogged, they were kept in bags and later transferred to plastic containers of water to maintain the wood structure essential for tree species identification. Allowing the wood to dry would have distorted the wood structure. Additionally, when the salt water evaporated, salt expanded within the wood and further distorted the wood structure. Wood samples were exported to the Archaeology Lab at Louisiana State

University (LSU) where they are housed in plastic containers filled with water. The time required for desalination varied with the density of the wood, the size of the sample, and the original site matrix. Once desalinated, the water in containers continued to be changed periodically to prevent slime and mold from developing (McKillop 2019).

Species identification of the waterlogged wood followed protocols of the USDA Forest Products Lab under the guidance of Dr. Mike Wiemann, who helped set up the lab at LSU for wood identification and established procedures that we followed. Each wood sample was cut using a backed razor blade on three planes, including the transverse, radial, and tangential, in order to be able to locate species-specific characteristics of tree species. The three cuts for each sample were placed on a microscope slide with a drop of glycerol to hold the sample in place and a slide cover was placed on top. The samples were examined under a microscope, with 100–200 × magnification useful for observing most wood structure. Microphotographs were taken for reference slides. Slides were initially grouped into categories with similar wood structures. An online identification key, "Inside Wood," was used to identify features characteristic of the wood structures of different species. Some species, such as rosewood or black mangrove, are readily identifiable because of unusual characteristics, such as "storied rays" and "ripped cells," respectively. Also helpful for identification were slides from modern wood samples as well as published microphotographs of wood.[1]

Tree Species Selection for Building Construction at the Paynes Creek Salt Works

Trees selected for building construction at the Paynes Creek Salt Works include palmetto palms (*Acoelorraphe wrightii*) and a variety of hardwoods (Figure 5.2). The 4,042 mapped architectural wood elements (posts, beams, and wedges) include 1,498 palmetto palms and 2,544 hardwoods. Of the total mapped wood, 2,468 architectural elements are identified to species, genus, or family. Identified elements include 1,498 from palmetto palms and 970 from hardwoods. The 970 hardwoods are vertical posts. Unidentified hardwoods include 32 horizontal beams, 73 wedges, and 191 vertical posts. In addition, 123 wood samples from Site 29 have not yet been examined. There are 1,155 vertical posts that were mapped but not sampled.

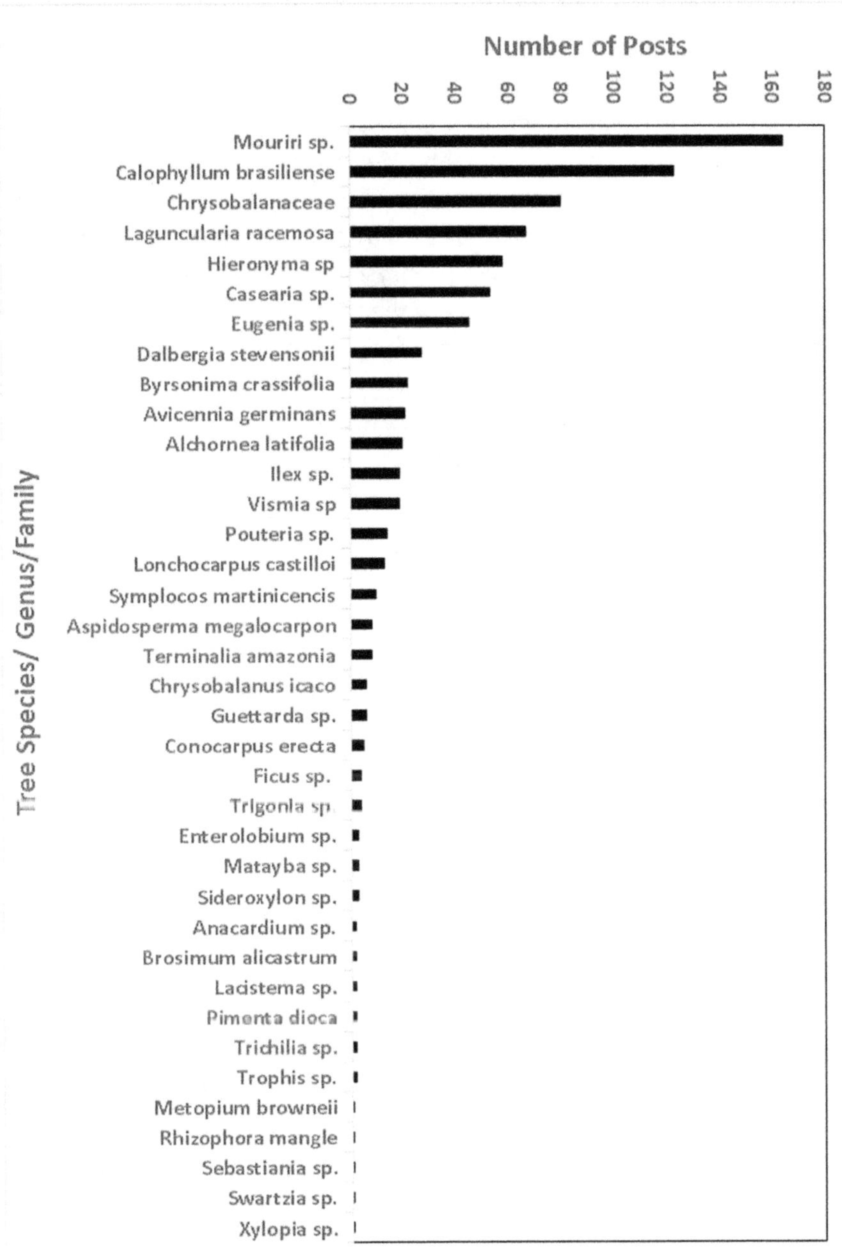

Figure 5.2. Tree species, genera, or families used in building construction at the Paynes Creek Salt Works.

In order to evaluate variability in wood use for construction of the salt kitchens, the sample of 970 identified hardwood posts was divided into two groups. The first groups include the 19 preferred woods—those with more than six posts per species, genus, or family. They are the main trees used in construction—notably, *Mouriri* sp., Santa Maria, Chrysobalanaceae, white mangrove, *Hieronyma* sp., *Eugenia* sp., and *Casearia* sp. (Figure 5.2). The wood was selected from the deciduous rainforest to the west (*Mouriri* sp., Santa Maria, Chrysobalanaceae, *Hieronyma*, *Eugenia*, *Alchornea*, and rosewood), from the mangroves immediately beside the salt works (black and white mangroves), and the coconut woodlands along the coast (*Casearia* sp., nance/craboo). Some of these trees are also available on the pine savanna, notably nance/craboo. Some vestiges of the relict vegetation and soils of the Classic period when sea level was lower and the Paynes Creek Salt Works were on land can be seen in the modern vegetation that includes some deciduous rainforest species in isolated stands in the mangrove ecosystem, notably, sapodilla and emery (*Vochysia hondurensis*). Neither was used in building construction. The K'ak' Naab' canoe paddle was made from sapodilla.

The second group includes the 24 less-desired woods that are represented by six or fewer posts per species (Figure 5.2). Trees used for six posts include *Trigonia* sp., glassy wood (*Guettarda* sp.), and icaco plum. Trees used for five posts include *Myrciaria* sp., fig (*Ficus* sp.), and buttonwood. The other species are represented by fewer posts. The value of the less common species is their association with other more common trees in different vegetation landscapes so that the travel patterns from the Paynes Creek Salt Works to acquire trees for building construction can be evaluated. The minor tree types are available in the same vegetation landscapes as the more common tree types. Buttonwood and red mangrove are in the mangrove ecosystem at the salt works. Icaco plums and cashew are available in the coconut woodland by the coast. Most of the other minor tree types are available in the deciduous rainforest to the south of the salt works, along Muschamp Creek and south of the Deep River (see Figure 1.2).

There were preferences for some woods for building construction in different areas of the salt works (Figure 5.3). Santa Maria is the first or second most popular wood in all areas. *Mouriri* sp. is first or third most popular wood except in the Stingray area, where it is sixth. White mangrove is third or fourth except in the West Lagoon, where it is sixth. Rosewood

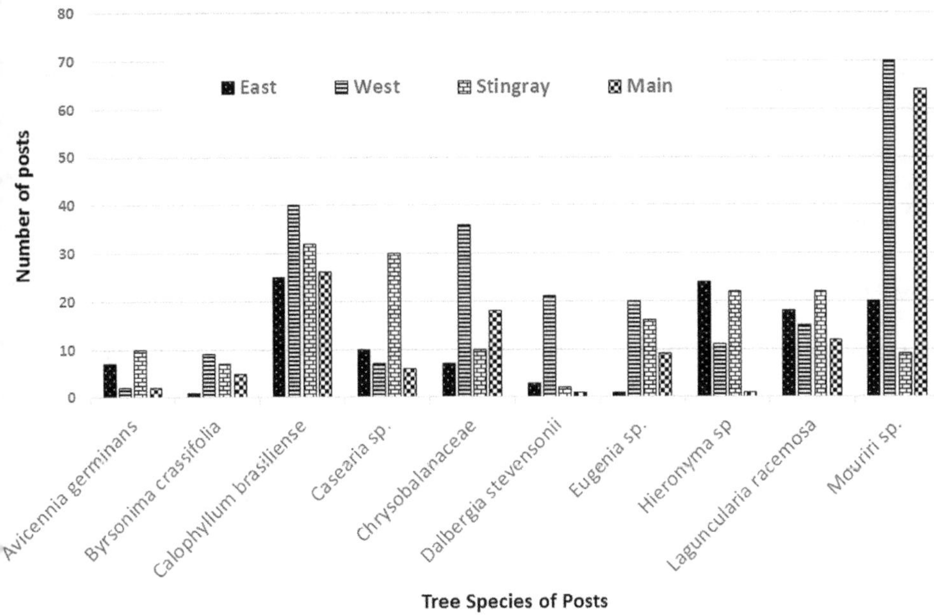

Figure 5.3. Bar graph showing preferences for some woods in different areas of the salt works. The West Point sites are excluded since wood posts were not sampled from that area.

is fourth in the West Lagoon and not a preferred wood elsewhere. Black mangrove ranks sixth in the East Lagoon and fifth in the Stingray area. A chi-square statistic of a two-way contingency table shows there is a significant relationship between wood type and area of the lagoon (Table 5.2).

All tree species in the sample were locally available, but not all common local tree species were selected for building construction. The mangrove ecosystem setting of the Paynes Creek Salt Works meant that red, black, and white mangroves along with buttonwood were the closest tree species. Pine, oak, and palmetto palms were available nearby on the savanna, locally known as "broken pine ridge" on the north and west sides of the lagoon. Deciduous forest abuts the lagoon system on the south, providing access to a variety of hardwood tree species (Wright et al. 1959; Figure 5.1).

Based on availability, some common species were absent or rarely used in building construction. Of the mangroves, only white mangrove was commonly used for posts, with one red mangrove post and black mangrove representing 3 percent of posts. Notable in absence is pine, available

Table 5.2. Comparison of most common tree species and area of the lagoon

	East Lagoon	West Lagoon	Stingray Area	Main Channel	Row totals
Calophyllum sp.	25 (20.63)	40 (43.32)	32 (26.56)	26 (32.49)	123
	0.93	0.25	1.11	1.30	
Casearia sp.	10 (8.89)	7 (18.67)	30 (11.44)	6 (14.00)	53
	0.14	7.29	30.09	4.57	
Chrysobalanaceae	7 (11.91)	36 (25.01)	10 (15.33)	18 (18.75)	71
	2.02	4.83	1.85	0.03	
Laguncularia sp.	18 (11.24)	15 (23.60)	22 (14.47)	12 (17.70)	67
	4.07	3.13	3.92	1.83	
Mouriri sp.	20 (27.34)	70 (57.41)	9 (35.30)	64 (43.06)	163
	1.97	2.76	19.50	10.19	
Column totals	80	168	103	126	477 (grand total)

Note: Each cell shows the observed value, followed by the expected in parentheses, and the X^2 per cell. $X^2 = 101.7949$, significant at $p < 0.05$.

locally on the pine savanna. Some species were preferred, notably Santa Maria and rosewood as well as *Mouriri* sp., *Hieronyma* sp., and *Casearia* sp. Palmetto palms were mainly used for lines of posts. Sapodilla and rosewood were used for objects, calabash for making containers from the tree's gourds, and trees with edible fruits were collected for food, including native palms (cohune and coyol), craboo, and mamey apple. Calabash was not used for posts. Apart from the native palmetto palms, the cohune and coyol palms were not used for construction although the endocarps are commonly found at the underwater sites.

Selection of Tree Species for Different Parts of a Building

Modern Maya use different wood for various parts of buildings, such as corner posts and roof beams (Redfield and Villa Rojas 1962:35; Wauchope 1938), so the occurrence of different kinds of wood in construction at the Paynes Creek sites is expected. Based on reports of modern and historic Maya buildings made of pole and thatch, the buildings at the Paynes Creek Salt Works were expected to show selection of desired local trees, with specific tree species preferred for different parts of buildings, such as corner posts.

Table 5.3. Tree species used for different parts of buildings at selected sites

Tree species	Corner Post	Wall support post	Interior support post
Mouriri sp.	Sites 23, 53, 58, 72	Sites 53, 57	
Rosewood	Sites 57, 72, 74	Sites 74, 75	Site 72
Santa Maria	Sites 32, 53, 57, 58, 72, 75, 77	Sites 57, 58, 74, 77	Site 74
Chrysobalanaceae	Sites 74, 77	Site 32	Site 74
Hieronyma sp.	Sites 32, 56	Site 32	
Casearia sp.	Sites 23, 57, 75	Site 75	
White mangrove	Sites 74, 77	Sites 57, 74, 75, 77	Site 77
Nance/craboo	Sites 74, 77		Site 77
Eugenia sp.	Site 77	Site 77	Sites 74, 77

Some woods were preferred for different parts of buildings. To identify patterns of tree species selection for building construction, I focused on common woods used at salt works with clear rectangular buildings, including Sites 74, 75, and 77 in the West Lagoon (Figure 5.4). Using those sites as a model, similar patterns of wood choice were noticed at other sites by plotting the location of the most common wood species (Table 5.3). For the well-defined rectangular buildings with posts identified to species, wood commonly used for corner posts includes Santa Maria, *Mouriri* sp., rosewood, Chrysobalanaceae, craboo, *Casearia* sp., *Eugenia* sp., and white mangrove. All but craboo also were commonly used as wall support posts.

From the patterns of tree species used at Sites 74, 75, and 77, post use can be inferred at other sites where rectangular building footprints are not as clearly defined (Figure 5.5). For example, Site 72 in the West Lagoon has four corner posts (Santa Maria, *Mouriri* sp., rosewood) with a center line of six rosewood posts that form part of one building. Site 23 in the East Lagoon has a clear pattern of using Santa Maria for lines of posts that may include exterior wall supports and roof overhang. Other wall supports include *Mouriri* sp., *Casearia* sp., black mangrove, and white mangrove posts. Corner posts include *Mouriri* sp., *Casearia* sp., and possibly Santa Maria. Site 32 has *Hieronyma* sp. for corner and wall supports, Santa Maria for a corner post, and Chrysobalanaceae as a wall post. Site 60 extends from the East Lagoon into the Main Channel and is the only

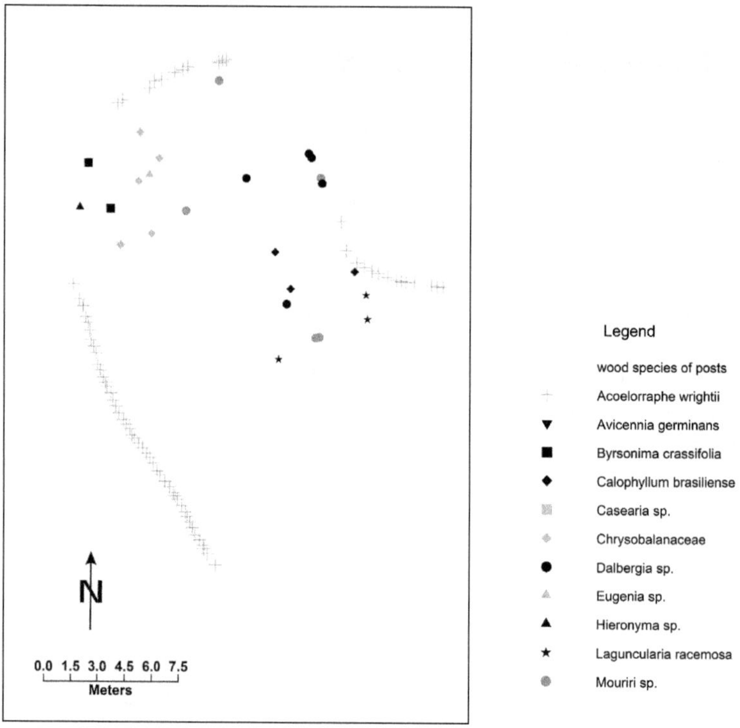

(a)

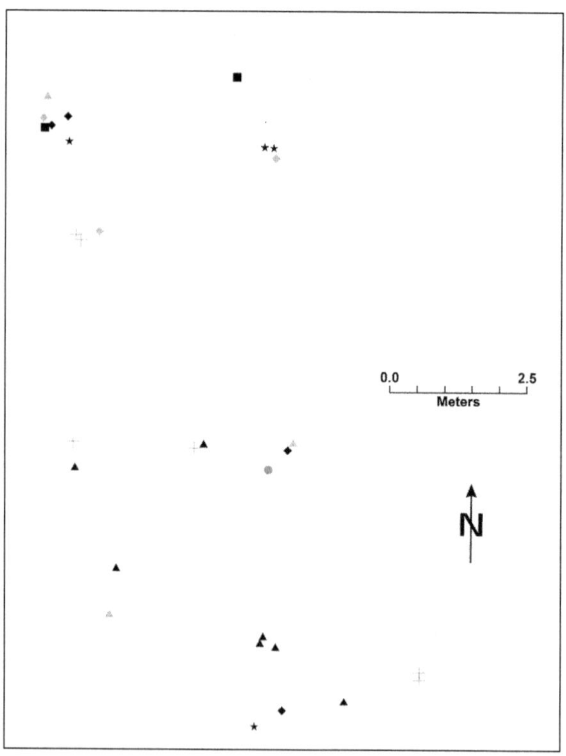

(b)

Figure 5.4. Tree species selection for Sites 74 (*a*), 77 (*b*), and 75 (*c*).

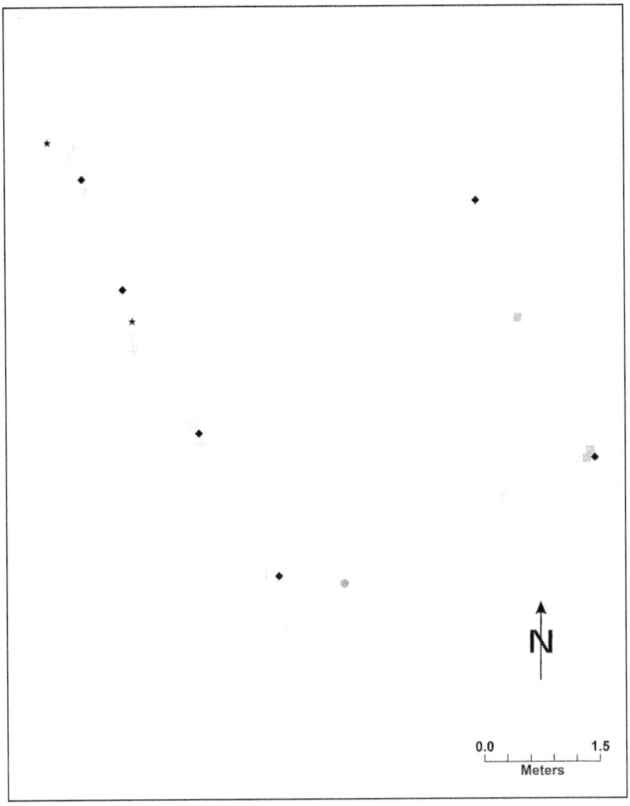

(c)

Main Channel site for which there are wood posts identified to species (Figure 5.6). *Mouriri* sp. is a common tree used for building construction at Site 60, along with Santa Maria. The orientation of the lines of posts resembles that at other sites (Figure 4.13). Many buildings are represented.

Summary

Tree species selection for building construction favored hardwoods from the deciduous forest (Santa Maria, rosewood), from the coconut woodland (nance/craboo, Chrysobalanaceae, *Casearia* sp.) and from the mangrove ecosystem (black and white mangroves). Trees were not selected from the broken pine ridge (pine savanna) where there were oak, pine, calabash, and sandpaper trees. The retaining walls at the edges of sites were from palmetto palms available on dry land inland from the mangroves about 100 m from the lagoon water's edge. Corner posts and supporting central

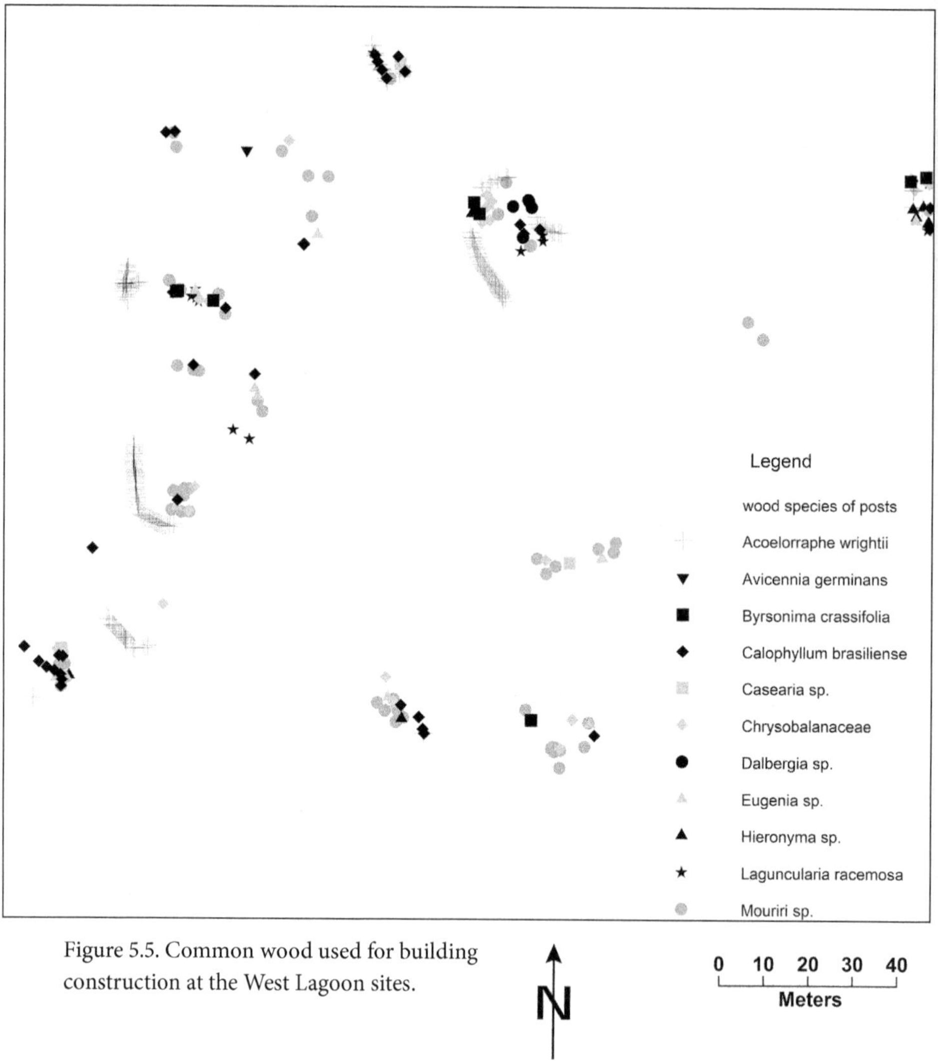

Figure 5.5. Common wood used for building construction at the West Lagoon sites.

posts were selected from hardwoods, with variation among sites, even buildings at the same site. Building posts and roofing thatch needed periodic replacement due to decay and termites. Insects inside wood post samples point to insect infestation as a common issue. Ethnographic data indicate post and thatch replacement every 10–20 years.

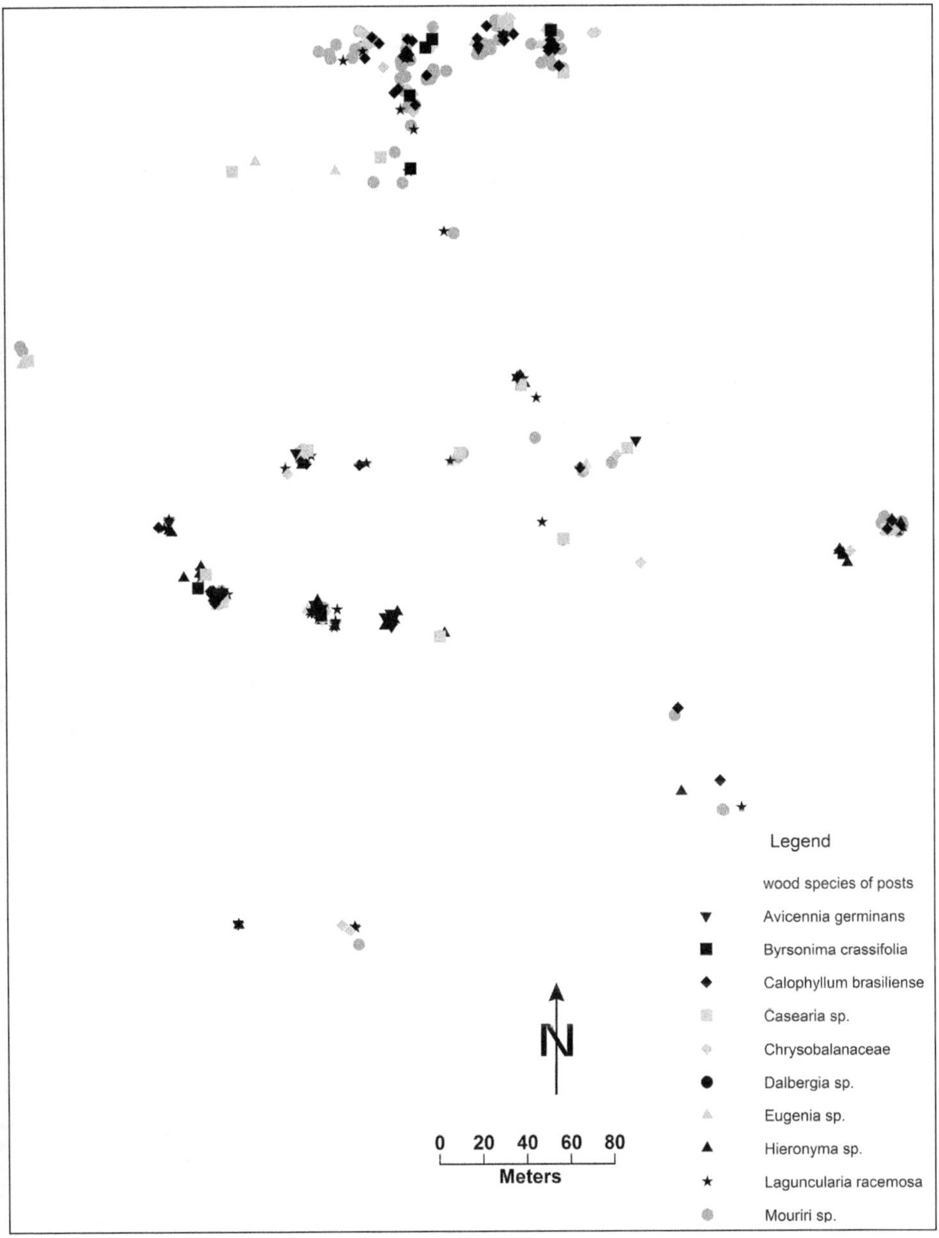

Figure 5.6. Common wood used for building construction at the East Lagoon sites.

6

Labor, Production, and Identity

Salt production required a labor force as well as a variety of supporting activities. The labor force may have included seasonal workers from inland or coastal communities, daily workers from nearby coastal communities, or permanent residents. The constitution of labor may have been household production or surplus household production. Dean Arnold (2017) describes modern pottery production in the Yucatán where members of households have separate workshops and employ people outside the household. Outsourcing allows specialization when there are multiple specialized tasks to be done. At Sacapulas, the salt kitchens are buildings where all activities are focused on salt production. In what Kenneth Hirth (2009a) refers to as "contingent multi-crafting," there were activities necessary for salt production at Sacapulas that took place elsewhere (Reina and Monaghan 1981). They included gathering firewood, making the clay evaporation bowls at home, enriching the salty playa soil near the hot springs, carrying the playa soil in baskets with tumplines up the hill to the leaching stations beside the salt kitchens, pouring water through salty soil in wooden boxes near the salt kitchens to enrich the brine before it was evaporated, using calabash bowls to scoop brine into jars inside the salt kitchens, acquiring the jars, and selling the salt twice a week at markets in different communities. In Thailand, the Philippines, and elsewhere, making salt requires many other contingent activities (Parsons 2001; Yankowski 2010; Yankowski and Kerdsap 2013; Yankowski et al. 2015).

Who Were the Salt Workers?

The salt workers' identity—if they were permanent residents, lived nearby at other coastal communities, or were seasonal workers from inland communities—is important for evaluating labor in ancient economies and how production was integrated with distribution. Placing overseers at salt works in ancient China (Flad 2013) contrasts with producers taking salt to markets themselves as at Sacapulas, middlemen traders such as Zinacantán merchants buying salt from the Ixtápa salt works and then selling the commodity elsewhere, or seasonal workers from inland communities who move to the coast to make salt for themselves (Andrews 1983:60; Antonites 2016; Reina and Monaghan 1981; Williams 2003). These factors are critical to evaluating the impact of the supply of salt to inland Maya where salt was scarce.

Peoples' residences are basic to defining identity where households are also the center of production. Residences have diverse activities, including cooking, eating, sleeping, maintenance and storage of objects and tools used in daily life, and socializing, among others. Activities at salt kitchens are focused on the single activity of producing salt. The evidence presented here indicates that the production units at the Paynes Creek Salt Works developed from surplus household production. Other researchers have found that household production was not restricted to the physical residence but also included other places where resources were extracted, such as jadeite outcrops or good farmland (Nietschmann 1973; Rochette 2009). Residences for salt workers elsewhere were located near sources of salty soil, salt springs, or the sea (Antonites 2016; Halliwell et al. 2016; Reina and Monaghan 1981; Yankowski 2010).

Did the Paynes Creek salt workers live near the source of super-saline salt on the shores of the shallow waters of Punta Ycacos Lagoon? Alternatively, did the salt workers live nearby, perhaps a short canoe paddle from Wild Cane Cay or another nearby coastal community? Did inland workers reside seasonally at the salt works or nearby on the coast? The concentration of the salt works in the lower reaches of Punta Ycacos Lagoon closest to the sea points to proximity of the sea for boat travel for canoe transport of salt but perhaps also for commuting to work.

Examining case studies of salt production from around the world reveals variability in the location and seasonality of salt workers' residences. Salt makers often live in residences near salt works, either in the same community or nearby, or they migrate seasonally to salt works. At Sacapulas, the salt makers' residences are located behind the salt kitchens, which were uphill from the salt spring (Reina and Monaghan 1981). On the island of Bohol in the Philippines, salt kitchens were set up along the coast for brine enriching and boiling (Yankowski 2010). At Ban Non Wat in northeast Thailand, grass-walled huts were set up to keep out wind for brine processing, and the salt workers lived nearby (Halliwell et al. 2016). Salt kitchens were located near the town at Ixtápa (Andrews 1983:56). Rowan Flad (2011) suggests the salt workers at Zhongba on the Yellow River in China lived nearby. In southeast China in Shandong Province, there were salt works along the coast with worker houses that were located farther back from the coast (Flad 2013). Jeffrey Parsons (2001) notes that salt workers lived in Nexquipayac approximately 2 km from their salt works. In west Mexico, inland settlers visited the coastal salt works seasonally and set up temporary camps nearby (Williams 2003). Salt makers at Sacapulas made most salt in the dry season but also carried out salt making in the rainy season when salt was needed, using stored salty soil and brine.

Other salt work sites along the coast of Belize are described as settlements where salt production was one of many activities. At Northern River Lagoon and Saktunja, salt production and salting fish were carried out at the community (Masson and Mock 2004). Moho Cay was a trading port in the mouth of the Belize River, and salt making was one of many activities at the island community (McKillop 2004). Late Preclassic salt production was specialized in a workshop at Cerros, which was a coastal trading port with a resident community (Robertson 2016). Salt production took place at the community of Marco Gonzalez (MacPhail et al. 2017). Salt production at Wits Cah Ak'al did not take place in the residences (Aimers et al. 2016; Murata 2011). The location of the salt makers' residences remains to be identified at the Placencia Salt Works (Sills 2016, 2017), which is located in a similar lagoon setting north of the Paynes Creek Salt Works.

If the salt workers lived at the Paynes Creek Salt Works, either in separate houses (as at Sacapulas) or in the same building (where salt making was carried out as part of household activities and not for surplus production), we would expect to find artifactual record or sediment chemistry evidence of household activities such as those found in Maya residences at nearby Wild Cane Cay or elsewhere in the Maya area. Material remains of household activities can include middens with plant and animal food remains, burials associated with houses, a variety of pottery vessel forms for household use in storage, cooking, and serving as well as obsidian, chert, and ground stone tools. The mangrove peat matrix of the underwater sites is too acidic to preserve bone. However, we would expect multicrafting, including remains of salt production and domestic activities similar to the independent household chert workshops found in the Mopan Valley (Horowitz 2017) and at Chan (Hearth 2012).

If evidence of residential activity is lacking, then the salt workers lived farther away—either on the coast at Wild Cane Cay or another nearby coastal settlement. Workers from the coast or inland communities may have visited the Paynes Creek Salt Works seasonally and set up temporary camps. Activities at nonresidential salt works may have been solely focused on salt production or may have included contingent activities such as making the salt pots and cleaning or salting fish.

If the Paynes Creek Salt Works were exclusively for salt production for workers who lived elsewhere, the artifacts should be predominantly or exclusively briquctage. In contrast, if the workshops were part of household residences, there should be a more diverse artifact assemblage including pottery and other objects for household use. If household production was spatially separated as at Sacapulas, then we should find areas of salt production spatially separate from residences. The linear arrangement of some of the Paynes Creek sites may have been to situate the salt workshops adjacent to the salty water, with the residences behind or even farther away (see Figures 4.15–4.16). Depending on the scale of production, living in a salt workshop would be unpleasant because of the salt residue and the salt-making supplies. After the abandonment of the Paynes Creek Salt Works, salt was locally produced on a limited, small-scale basis for household use during the Postclassic period at the major trading port of Wild Cane Cay and at nearby Frenchman's Cay (McKillop 2002:112).

Looking for Residences at the Salt Works

Most of the sites were in five areas of the lagoon system (see Figure 4.7). There are two patterns of sites, including long lines of sites that may correspond to locations of ancient shorelines as well as clustered sites with buildings and plazas. Larger sites (7 and 60) with multiple buildings may have included residences and salt kitchens (McKillop 2018). The smaller sites may have focused on salt production, including brine enrichment, brine boiling, and storage. There are 405 vertical wooden posts at Site 7 demarcating the outlines of 10 buildings and three plazas (Figure 6.1). Site 60 likely has about 10 buildings, but the outlines of buildings are obscured by the high density of artifacts on the sea floor and the difficulty

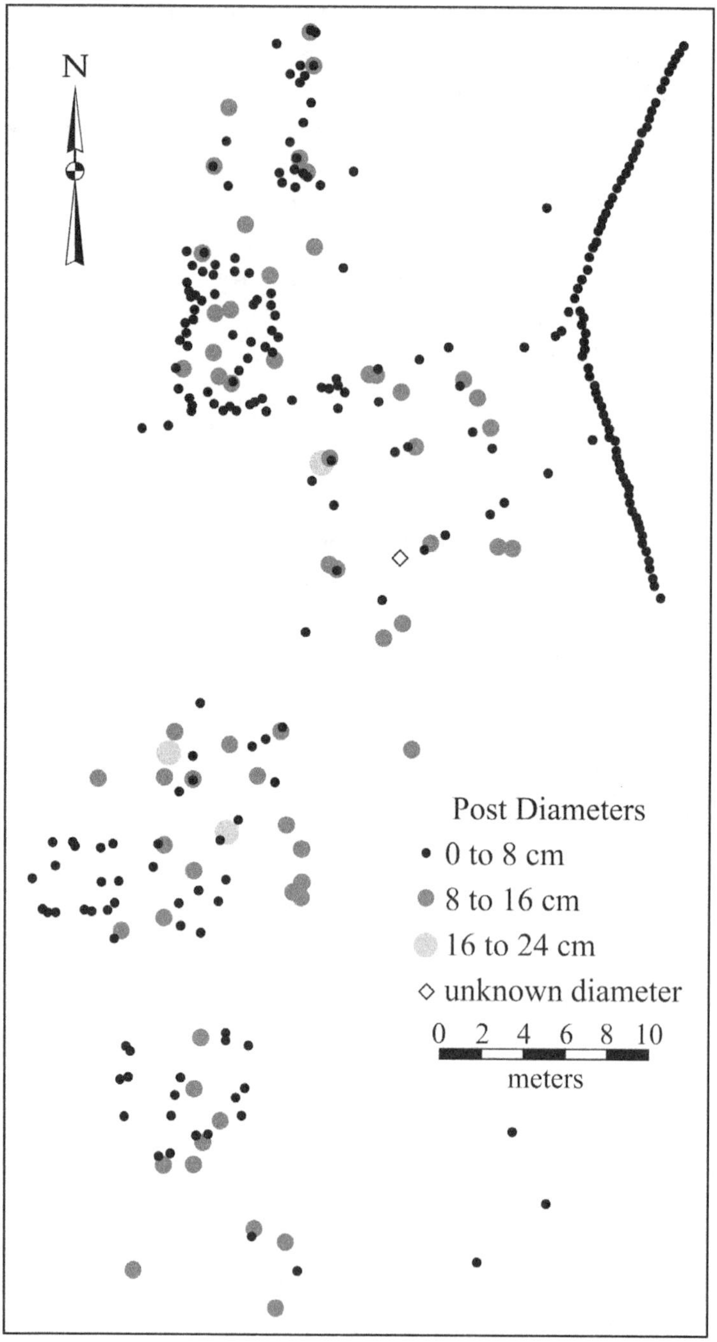

Figure 6.1. Post sizes at Site 7.

of flotation survey in the deeper parts of the underwater site (see Figure 4.13). These two large sites are located at the entrance to different areas of the salt works. Site 60 is at the entrance to the East Lagoon and Main Channel. Site 7 is located at the entrance to the West Point area.

Examination of the spatial patterns of artifacts helps to identify the organization of salt production and contingent activities. The map of artifacts at Site 7 shows most artifacts were along the inside walls of buildings (Figure 6.2). The distribution was not visible until the systematic flotation survey and flagging of posts and diagnostic artifacts was completed, mapped, downloaded to the laptop and displayed on the project GIS, Underwater Maya. Pottery mapped during flotation survey shows spatial patterning of different pottery types. The Punta Ycacos salt-making pottery is only associated with some wooden buildings. One building has Belize Red and Warrie Red pottery but no salt-making pottery, suggesting that different activities took place in different buildings and in the plazas. The lack of pottery in the three plazas underscores that the evaporation of brine was carried out indoors and that other activities took place outside. Artifactual evidence supporting non-salt-making activities include a vase with an incised glyph band below the rim, a small bottle, and a unit-stamped pot (Figure 6.3a–c). Other unusual pottery includes Terminal Classic, gouge-incised vessels from nearby Site 8 and from Site 61 in the Main Channel (Figure 6.3d–e). Unit-stamping is uncommon at Site 7. Large and small incense burners were mapped. There are no figurine whistles (ocarinas) at Site 7.

In contrast, figurine whistles and unit-stamped jars are common at Site 60. Various styles of ball players are depicted in the figurine whistles from Site 60 (Figures 6.4). The other main theme of figurine whistles is a woman, sometimes with a child or infant or holding something else. Warrie Red jars have a variety of unit-stamp decorations around the neck and shoulder, sometimes also extending farther down the body, as with an elaborate unit-stamped jar from the Stingray Lagoon site (McKillop 2002:Figure 3.27). Unit stamps of the S, reverse S, monkey, and comb stamping were concentrated in the Main Channel and in the adjacent East Lagoon. Site 60, which extends from the Main Channel and underneath mangroves to the East Lagoon, has many unit-stamped Warrie Red sherds.

Figure 6.2. Mapped pottery on the sea floor at Site 7 by artifact type.

(b)

Figure 6.3. Terminal Classic pottery. (*a*) vase sherd from Site 7A with incised glyph band below rim; (*b*) small bottle from Site 7G; (*c*) Warrie Red jar with abstract monkey motif, circles, lines, and punctate from Site 7E; (*d*) gouge-incised pottery from Site 8A; (*e*) gouge-incised pottery from Site 61A. Drawings by Mary Lee Eggart.

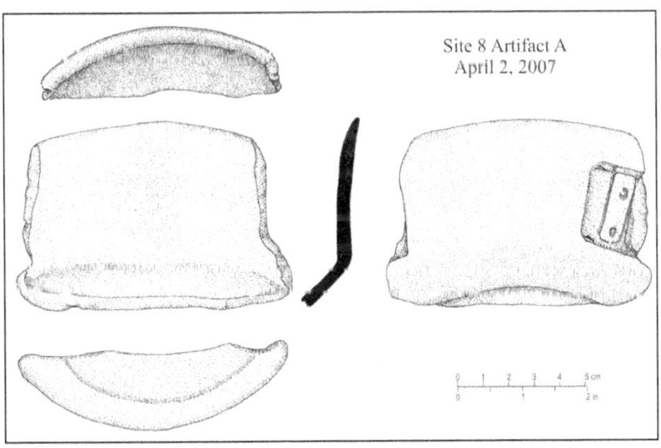

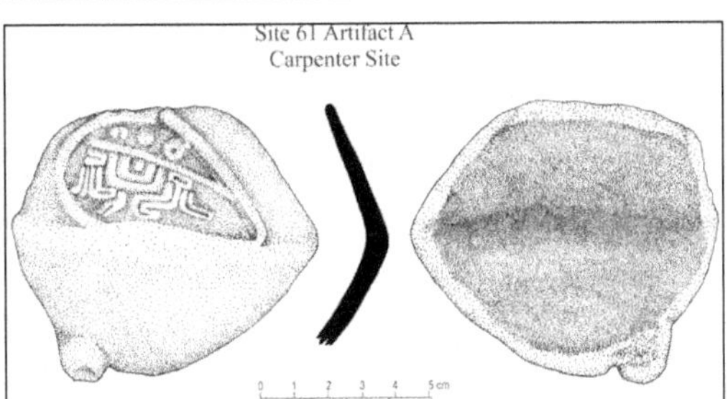

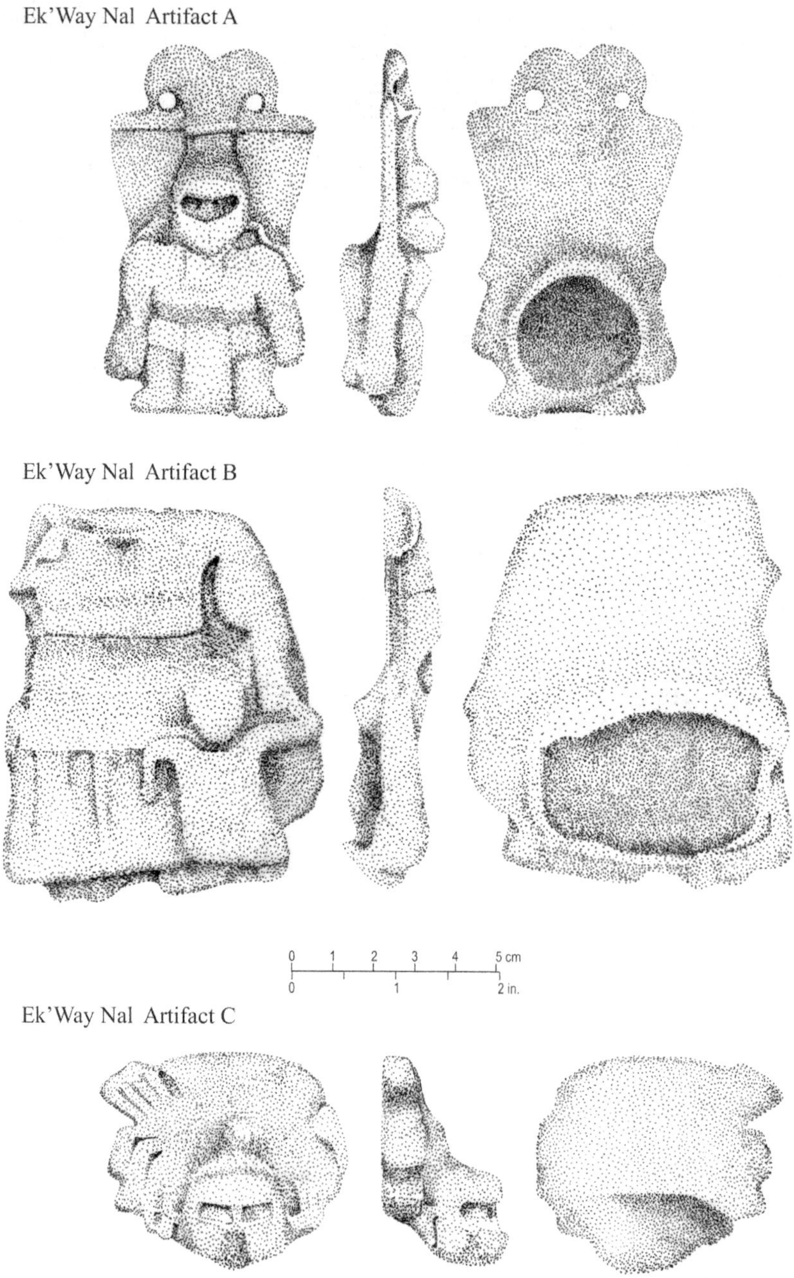

Figure 6.4. Figurine whistles (ocarinas) from Site 60 (Ek Way Nal) showing ball players and woman and child. By Mary Lee Eggart.

Table 6.1. Briquetage densities inside buildings at the Paynes Creek Salt Works

Site	Briquetage density (% of pottery by weight)
74	98.4
58	97
35	89
29	94
24	90
23	91
11	95

Densities of Briquetage

The density of briquetage in transect excavations was used as a gauge of the effort put toward salt production. Briquetage densities were high, consisting of 89–98 percent of the pottery as calculated from material excavated along the inside walls of buildings at several sites (Table 6.1). Salt production was the main activity at the salt works along ancient shorelines. The number of vessels per square meter was calculated using pottery rims from Site 74, Transect 1, including all 14 units and levels to 60 cm depth. The transect was primarily inside Building A at the site but also extended 2 m beyond at each end (McKillop and Sills 2016). The results are 9.9 Punta Ycacos salt rims, 0.3 Warrie Red water jars, 0.8 Mangrove Unslipped water jars, and 0.2 Belize Red serving bowls. The Warrie Red and Mangrove Unslipped jars were used to store brine and loose salt. Excluding the Warrie Red and Mangrove Unslipped pottery, that leaves two Belize Red body sherds that were not part of salt production. Ceramics show that salt production was the dominant activity.

Contingent Multicrafting at the Salt Works

Specialization to meet inland demand for salt may have resulted in independent work units consisting of householders and others, whose energy was focused on salt production and its various associated contingent, multicrafting activities—the manufacture of products that are used in the main crafting process. The production unit likely consisted of household members as well as others outsourced to fulfill specific tasks such as building canoes, making pottery, or transporting salt to market. Activities related to salt production at the Paynes Creek Salt Works included making salt pots, cutting trees and making posts that were transported to salt works to make buildings, gathering firewood, woodworking, enriching brine, salting fish, crafting calabash bowls and other objects, transporting salt and salted fish to inland and coastal consumers, and making canoes and paddles. Lumbering, construction, and woodworking to support the infrastructure of salt production were significant activities at the Paynes Creek Salt Works. Trees were felled, transported, and stored. Posts and beams were cut for building construction. Thatch was cut, transported, and used in roofing.

Flotation survey and mapping of artifacts on the sea floor yielded artifacts, wood, and marine sediment to examine variability in activities, including possible residences for salt workers or overseers. Due to the preservation of wood, research can focus on discovering variability in activities within and outside buildings. The use of sediment chemistry helps identify activity areas, especially since the acidic peat matrix of the site preserves wood, stone, and pottery but not bone. Use-wear analysis of stone tools is used to identify activities involving stone tools. Petrographic analysis is used to identify the origin of pottery vessels used in salt production and other minor pottery types.

The various activities contingent to salt production are discussed below. There were a variety of activities that took place on a regular basis. Some of the contingent tasks may have been done by the salt workers themselves, as with the Sacapulas salt workers making the salt pots at home and bringing them to their salt kitchens. Some activities may have been outsourced to relatives or others, such as making salt pottery, enriching the brine, building and refurbishing salt kitchens, cleaning and salting fish, fishing, woodworking, and transporting salt and other commodities to market by canoe.

Making Pottery Used in Salt Production

The salt workers or potters made salt pots on a regular basis from locally available clay and quartz sand. The salt pots are friable, and the clay is poorly mixed and fired at a low temperature, all factors that indicate the pots were not made for transport, although they could have been transports with hardened salt inside. Some buildings, perhaps residences, may

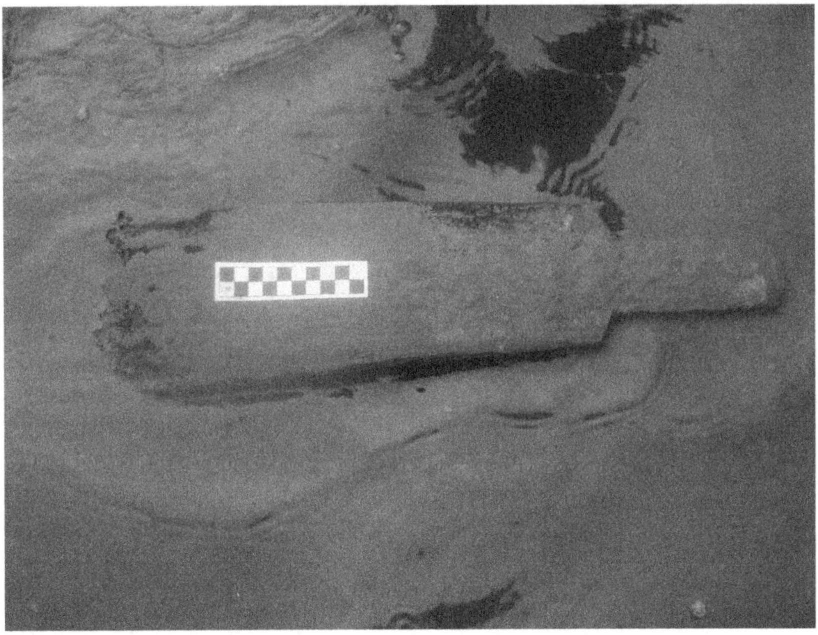

Figure 6.5. Photo of wooden pottery-making paddle from Site 103 after discovery.

have been used for making salt pottery, whereas others were used for making salt.

A wooden pottery-making paddle was discovered during flotation survey at Site 103 (Figure 6.5). One end of the paddle extruded from the sea floor and was worm-eaten. The portion below the sea floor was preserved, albeit salt-waterlogged. The Site 103 pottery paddle is identical those used by modern potters, including traditional potters who make salt-making pottery in northeast Thailand (Halliwell et al. 2016:Figure 5.7). A pottery paddle is used with a smooth stone or anvil to form, thin, and smooth pottery vessels. The stone is held inside the vessel, while the potter hits the exterior with the blade or handle of the pottery paddle. Pottery paddles often are cord-wrapped or carved to create a design on the pottery vessel. The Site 103 pottery paddle was plain, although a resin that covered most of the blade may be remnants of cloth or other material that was adhered to the wood to make designs. This possibility makes sense since the salt-making pottery—Punta Ycacos Unslipped jars, bowls, and basins—were rough on the exterior, which facilitates handling them. The bodies of the salt-making pots were thin, a trait useful for conducting heat in brine boiling, so using a pottery paddle may have facilitated thinning the pottery vessel walls.

In Thailand some pottery paddles were heirlooms passed down through a family, but sometimes male and female potters preferred to make their own paddles. In Thailand, there were some communities where men made the pots, pottery paddles, and selected the wood (Halliwell et al. 2016:129–130). The pottery paddle from Site 103 was made from rosewood (*Dalbergia stevensonii*), which was the commonly used wood selected for small objects at the salt works. Site 103 may have been a location for pottery making, as there is abundant charcoal and a distinctive pattern of wooden posts not found at other sites. Lines of palmetto palm posts define two walls of a building or area, with a large open area. Briquetage is present as well as unit-stamped pottery. Future excavations may clarify the ancient activities at the site.

Fishing or Salting Fish

Salting fish was evidently a major activity at the salt works, despite the virtual lack of fish bones or other bones. A vertebra from a large, unidentified fish and a manatee rib bone fragment were preserved in thick silt above the mangrove peat sea floor at Stingray Lagoon (Site 56). Two unidentified animal bones were preserved in a shell midden at the Eleanor Betty site (Site 67). Artifactual evidence of fishing includes a side-notched stone fishing weight from Site 7 (Figure 6.6) and a partially perforated pumice disk from Orlando's site that may have been intended as a fishing float (McKillop 2002:Figure 3.42a). In contrast, notched fishing weights made from sherds were common in midden deposits at Wild Cane Cay. A formed clay fish made from local sand and clay was found at Site 31 in the East Lagoon.

In contrast to the limited evidence of fish bones or fishing equipment, use-wear analysis of chert stone tools by Kazuo Aoyama (McKillop and Aoyama 2018) indicates most of the chert tools were used to cut fish or meat, with some used for scraping hides or scaling fish (Figure 6.7). Some of the tools also were used to cut or whittle wood. The overwhelming use of chert tools was for cutting fish or meat. The lack of fish bones may simply be due to the acidic matrix of the sites. However, the fish may have been cleaned and salted at the salt works and then transported whole to inland markets. Fishing would have been a time-consuming activity,

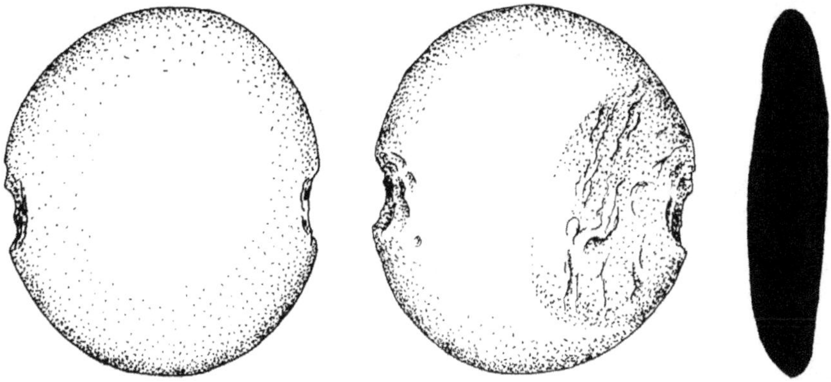

Figure 6.6. Side-notched stone fishing weight mapped from Site 7 (GGG). By Mary Lee Eggart.

conflicting with salt production. The lagoon system is a nursery for various fish as well as a home for manatees, sharks, crocodiles, barracudas, bonefish, goliath grouper, and permit. More fish are located in nearby Port Honduras, with abundant fish bones recovered from Classic period middens at Wild Cane Cay (McKillop 2005b:36–37). If the salt workers lived there, they may have obtained fish for salting from fisher folk on the island, traveling the short distance to the salt works to kill, cut, and salt fish (McKillop and Aoyama 2018). At the nearby inland city of Lubaantun, 39 percent of the animal remains were marine fish (Wing 1975), including species found in Late to Terminal Classic middens at Wild Cane Cay (McKillop 2005b:36–37) and available in the coastal waters near the salt works. Underrepresentation of estuarine catfish vertebrae in comparison to catfish heads at Northern River Lagoon suggested to Marilyn Masson (2004:114) that catfish were salted for inland trade. Cut marks on the ventral surface of tuna vertebrae at Colson Point indicates the fish were cut and splayed for drying (Graham 1994).

Fish sauce and fish products may have been the most important trade commodity after wine, olive oil, and grain in Roman times (Curtis 1984:58–59, 74–75; Locker 2007; Yankowski et al. 2015:9). In Thailand, China, and the Philippines, salting fish was important, but salting fish to make fish sauce provided a critical food that could be stored for later use and traded near and far (Flad 2011; Halliwell et al. 2016; Yankowski 2010). Salt-drying fish is preferable to sun drying since the fish lasts longer, especially if the fish is fermented (Yankowski et al. 2015:9).

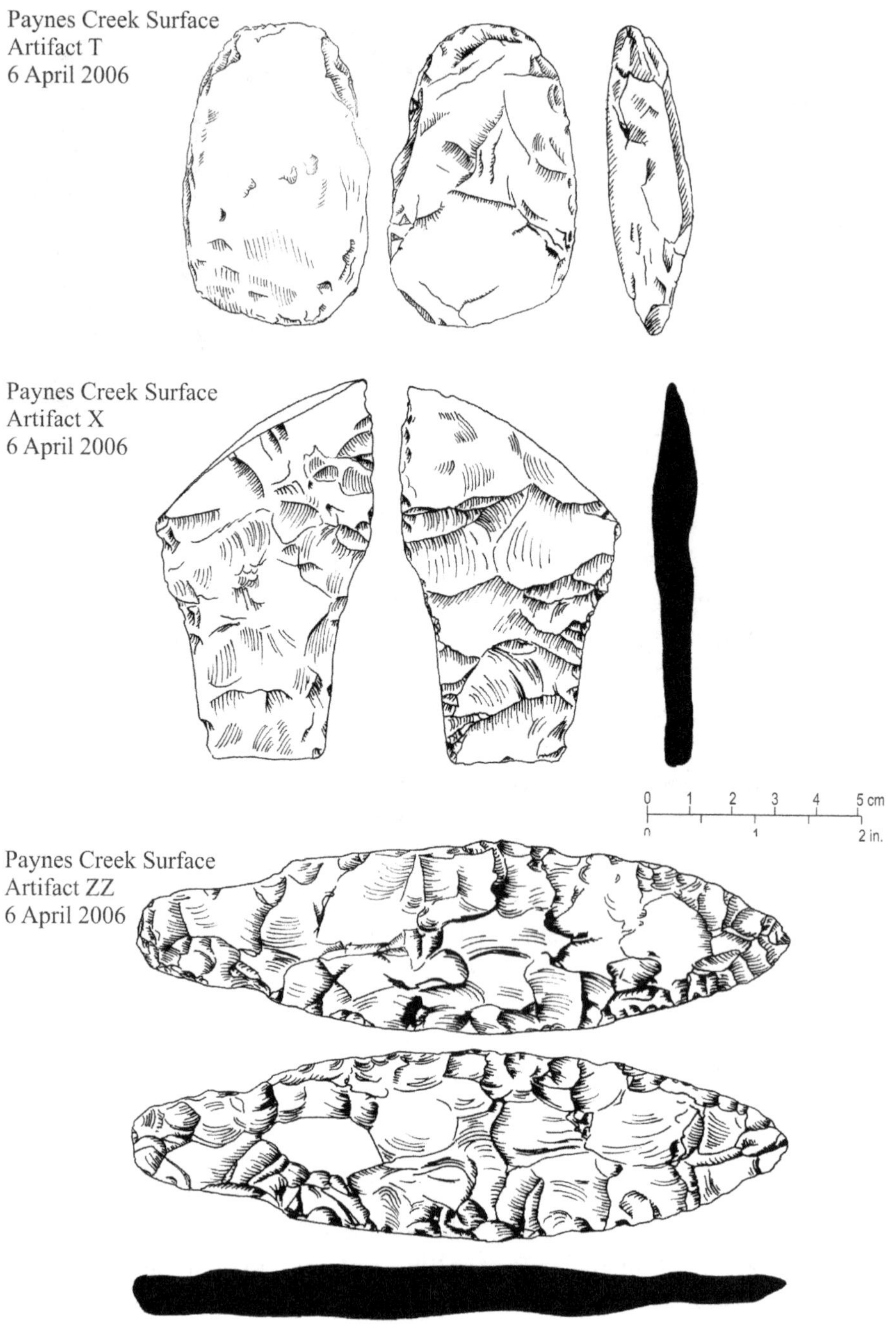

Figure 6.7. Chert tools from Ek Way Nal (Site 60) used for scraping hides (60T); cutting fish or meat (60X, 60ZZ), from use-wear analysis by Kazuo Aoyama (McKillop and Aoyama 2018). By Mary Lee Eggart.

Soil chemistry has been used in Alaska to detect areas of fish processing (Knudson and Frink 2010). Chemical analyses of marine sediment from the Chan b'i site at the Paynes Creek Salt Works revealed concentrations of calcium outside the salt kitchens near some modern mangroves (Sills et al. 2016). There also were high levels of phosphorus in and around the salt kitchens that were attributed to wood fires for the brine boiling, but the phosphorus also could indicate fish processing.

Food

In addition to fish, shells and plant foods were recovered (Figure 6.8; Feathers and McKillop 2018; Feathers et al. 2017; McKillop 2017a). Low in carbohydrates but high in calories and fat, the food items represented a variety of edible resources that may have supported a resident population or temporary or daily workers. They may have traded salt for corn at inland markets (see Eppich and Freidel 2015). They ate coyol and cohune fruits from native palms that were abundant in Site 74 excavations

Figure 6.8. Cohune endocarp interior view (*Orbignya cohune*), palm frond stem, mamey apple seed from Site 56, and burnt stick from Site 61D, interior view. Scale is 10 cm.

(McKillop and Sills 2016) and from flotation survey. Other plant food remains include craboo seeds (as well as the wood used in building construction) and mamey apple seeds. Wood used in building construction with edible fruits includes icaco plums (also called coco plums) and sapodilla fruits (the K'ak' Naab' canoe paddle was made from that wood).

The time of year that the fruits ripen indicates the seasonality of use of the salt works that does not correspond to the dry season, which ends mid-May in the area (see Figure 4.1). Instead, the seasonality of ripe tree fruits commonly recovered at the salt works indicates their use year-round. Mamey apples ripen in June on wild trees in the deciduous rainforest. Craboo ripen in late June through early August on large trees that grow in the pine savanna, coconut woodland, and deciduous rainforest. Icaco plums ripen in June through August on small trees in the coconut woodland and pine savanna. The coyol fruit is available twice a year, first in May through June and then again in November through December (Lentz 1990:189). The fruit and the nut inside are high in calories from both fat and protein. David Lentz estimates coyol could have supplied about 7 percent of Copan's caloric needs. Cohune fruits mature in December in Belize and are at their driest in May, at the height of the dry season in the area (McSweeney 1995:166). Coyol fruits are often gathered and eaten for the fleshy fruit around the endocarp. Both the coyol and cohune can be used to make oil, and the fruits can be stored to make oil at any time. Cohune oil was made in Belize historically and is a laborious task (McSweeney 1995:166–167) but may have been part of household subsistence activities in the Maya past (McKillop 1994, 1996b). Certainly, it was a common recovery in the excavations of Building A at Salt Work 74 (McKillop and Sills 2016) and was commonly found in household middens at Wild Cane Cay and other nearby cays (McKillop 1994, 1996b).

Building Construction

The salt workers selected trees for making wooden posts from a variety of nearby environments, sharpening one end of each post (Figure 6.9). Tree species were selected from the deciduous forest to the west, from the mangrove ecosystem itself, and from the coconut woodland along the coast, but probably not from the pine ridge to the north and west, as discussed further in Chapter 5. Wood procurement required boat transport

Figure 6.9. Sharpened end of wooden post 305 from Site 60 (Ek Way Nal), with bark preserved.

to the various locations and for hauling logs, either in the water or inside large, shallow-draft canoes. More Santa Maria, rosewood, and *Pouteria* sp. posts from the deciduous rainforest south of the Deep River were used for building construction in the West Lagoon than in the East Lagoon, which was farther away from the deciduous forest (see Figures 4.7 and 5.3). Proximity was not the only factor in tree selection since the mangrove forest species white and black mangrove were most popular in the Stingray Lagoon area, closest to the deciduous forest. In any event, trees were selected, felled, and transported using boats to the different areas of the salt works. A minority of stone tools were used to cut or whittle wood, as indicated by use-wear analysis by Kazuo Aoyama (Figure 6.10; McKillop and Aoyama 2018). Wooden posts with chipped ends were 3D imaged in order to have a research record of the chipping (McKillop 2016a).

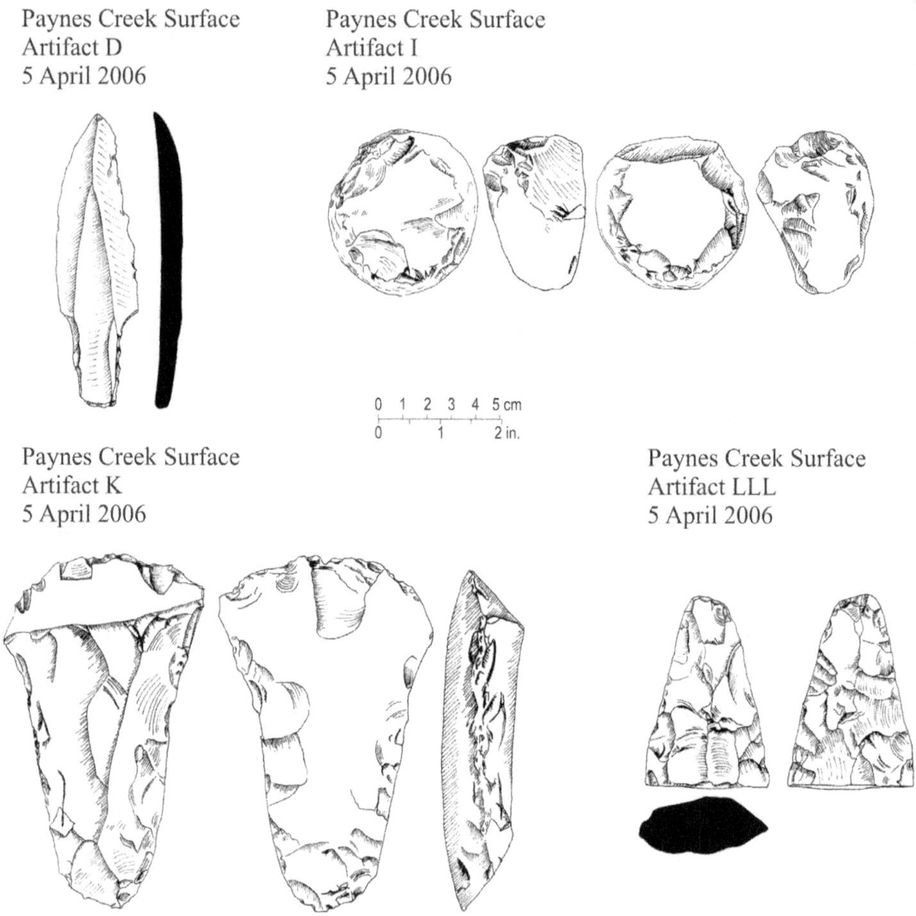

Figure 6.10. Use-wear analysis by Kazuo Aoyama shows some chert tools were used for cutting or whittling wood and cutting fish or meat (70D, 70K); whittling wood (70I); cutting fish or meat or scraping hides (72LLL; McKillop and Aoyama 2018). By Mary Lee Eggart.

Making Wooden Objects

Use-wear evidence of cutting and whittling wood indicates that woodworking took place at the salt works. In addition to the 4,042 wooden posts and other architectural wood, various wooden objects have been found on sea floor survey at the salt works. Woodworking workshops have not been identified although the chert tools were found in specific buildings at sites 60 and 72, suggesting there was more than one woodworking location. The fact that each building post was sharpened at one end to drive the post into the ground indicates that chert woodworking tools were widely distributed as part of building construction tool kits. Moreover, trees were felled elsewhere in the deciduous rainforest south of the Deep River, in the mangroves, coconut woodland, and pine savanna, underscoring the broad use of chert woodworking tools by the salt workers both at the salt kitchens and in various locations to select and fell trees. Young trees of standard diameter were selected and felled, demonstrated by the common presence of bark on wood posts and the uniform range of post diameters. Maintenance of the salt kitchens required periodic replacement of posts. This was noted at Orlando's site (Site 58), where a post was embedded in sand instead of the mangrove peat matrix. Palm thatch was replaced periodically, requiring travel to the deciduous rainforest where cohune palms were common.

Rosewood was the favored species for making small objects. A wooden handle identified as rosewood was found at the base of post 255 at Site 60 (Figure 6.11; McKillop et al. 2019).[1] The wooden pottery paddle from Site 103 also was made from rosewood, as were two paddle blade fragments from Sites 74 and 83. The K'ak' Naab' canoe paddle, in contrast, was made from sapodilla wood (*Manilkara zapota*). Other wood objects include an oval wooden mirror back from Site 12, a wooden rod with a hole from Site 83, and a triangular-shaped wooden object that once had a handle from Site 61 (Figure 6.12). The Site 61 object may have been a spoon for stirring the brine as it evaporated in pots over the fire, as indicated by modern ethnographic studies.

Making Canoes and Paddles

Water transportation was needed for obtaining clay and temper for making salt-production pottery, for obtaining wood for fuel and building construction, for transporting salt and salted fish to inland marketplaces, and for travel to residences if the salt workers did not live near the salt kitchens at the salt works. The K'ak' Naab' canoe paddle along with a smaller paddle from Site 7, blade fragments from Site 74 and 83, and the canoe from Site 67 are the only known physical evidence of canoe navigation among the ancient Maya (Figures 6.13–6.15; McKillop 2017b; McKillop et al. 2014). The canoe used in the salt-leaching process at Site 67 was

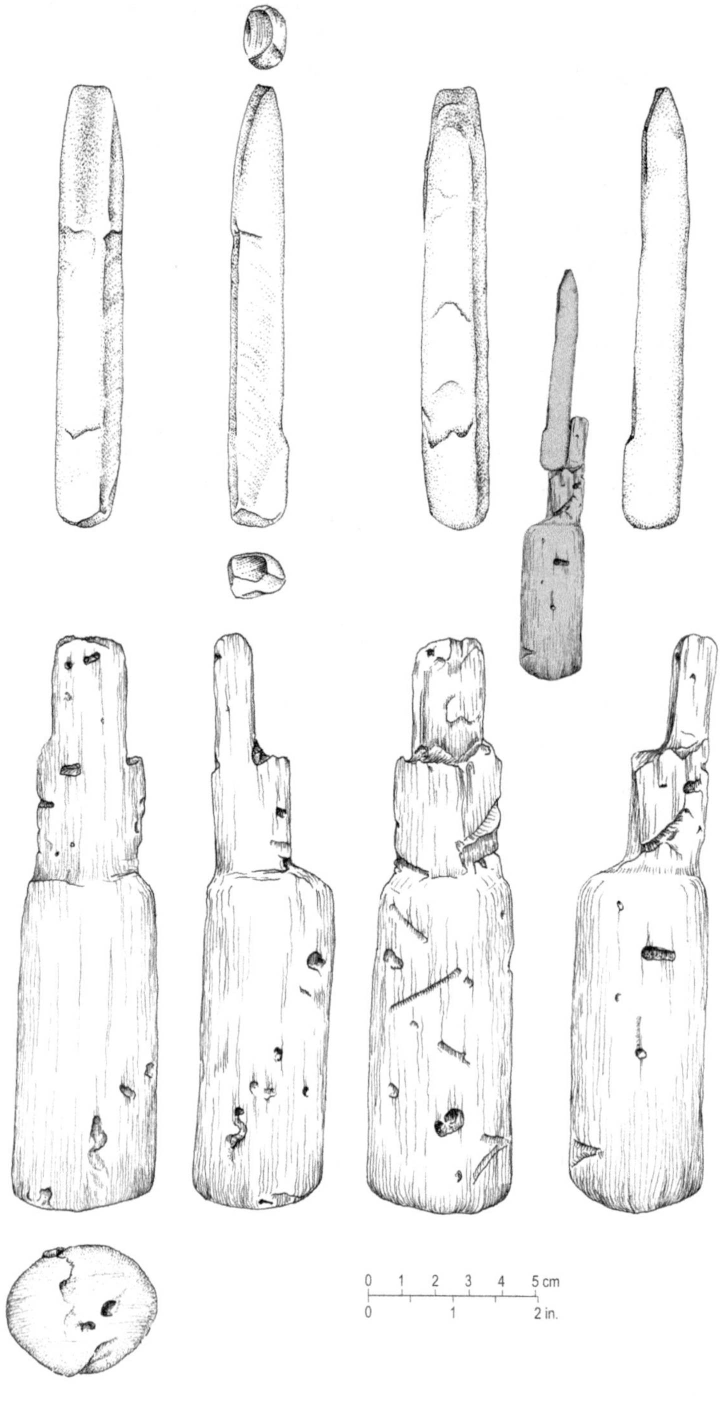

Figure 6.11. Rosewood handle and jadeite gouge from Ek Way Nal (Site 60). By Mary Lee Eggart.

Figure 6.12. Wooden spoon / paddle blade from Site 61F after conservation.

Figure 6.13. K'ak' Naab' canoe paddle after discovery. Line drawing shows paddler gods depicted on bone carving from burial 116, Temple 1, Tikal. By Mary Lee Eggart.

Figure 6.14. Wooden canoe paddle from Site 7 after discovery.

Figure 6.15. Wooden canoe paddle blade fragment from Site 83 after conservation.

Figure 6.16. Clay boat model from Site 74 after discovery. Artifact is 8 cm in length.

not made at that location, as indicated by the amount of chipping debris associated with manufacture of a dug-out canoe. A canoe model made from local clay and sand was found at Site 74, imaged at the base camp at Village Farm, with 3D replicas printed in ABS+ plastic at the LSU Digital Imaging and Visualization in Archaeology Lab (Figure 6.16; McKillop 2016a). Other canoe models made from clay were found at Orlando's site and Stingray Lagoon site and, carved from manatee rib bones, at Moho Cay (McKillop 2002:Figure 3.38b). A canoe model carved from manatee rib bone was also found at Altun Ha (Pendergast 1979:Figure 46b).

Wood Fuel

Fuel wood was cut, stored, and dried—likely indoors, as at Sacapulas (Reina and Monaghan 1981)—for evaporation by heating brine in pots over fires in the salt sheds. Wood fuel was locally available in the vicinity of the Paynes Creek Salt Works, with a variety of hardwoods in patches of dry land by the mangroves as well as deciduous rainforest to the south along the Deep River (Wright et al. 1959). Charcoal identified from the salt works indicates a variety of woods were used (Robinson and McKillop 2014). The pine savanna abutting the northern shores of the saltwater lagoon system hosts scattered clumps of pine, oak, and craboo.

Sediment coring indicates the mangrove ecosystem that dominates the modern landscape has persisted since the Classic period salt works, so mangrove wood—dense hardwood used elsewhere for charcoal—would have been available as well. A few charred cohune endocarps were discovered in the masses of charcoal, indicating that cohune fruits were not a major source of fuel although they are known to be a good charcoal and would have been good for a long, hot fire for boiling brine in the salt kitchens (McKillop 1994, 1996b; McSweeney 1995). Uncharred cohune and coyol endocarps were abundant in Building A at Site 74, along with remains of other foods, including mamey apple and craboo (McKillop and Sills 2016). One coyol endocarp from Site 67 was carved into a bell.

Calabash Carving

Other craft working included making calabash (*Crescentia cujete*) objects, including several calabash bowls. Calabash bowls were found at Sites 56 and 67, the latter the location of the canoe used for brine enrichment. The calabash bowls may have been used to move brine from large storage jars to the brine boiling pots over the fire.

Spinning Cotton

Spindle whorls were not common at the salt works. A spindle whorl from Site 7 flotation survey has use-wear marks from spinning material, likely cotton, which is cultivated on the coast today. Cotton may have been used in industrial purposes for making fishing lines or for clothing. A perforated potsherd disk from Site 61 has uneven edges (Figure 6.17). A notched stone pebble from Site 7 was the only net sinker found at the salt works (Figure 6.6). Cotton cloth was a tribute item as well as a currency equivalent. Surplus household production of cotton at Xuenkal, a community located midway between Chichén Itzá and the north coast of the Yucatán, was indicated by unexpected high number of spindle whorls (Ardren et al. 2016). Evidently cloth was produced as a tribute item for nearby Chichén Itzá. Joanne Baron (2018a:100) suggests that during the Late Classic in the Maya area, "textiles underwent a process of monetization that saw them take on the roles of change medium, measure of value, means of payment, and store of wealth." Although I am not suggesting that cotton was spun and woven to make cloth as a currency equivalency

(a)

(b)

Figure 6.17. Perforated potsherd disk from Site 6NN, exterior (*a*) and interior (*b*) views.

at the Paynes Creek Salt Works, the salt makers were working in an economic landscape featuring surplus production, marketplace trade, and accumulation of wealth requiring standard units of exchange (Freidel et al. 2016). Cotton, salt, cacao, and other commodities were traded in the marketplace at cities, towns, and villages, as dramatically visualized by the painted images of vendors and their wares on the exterior of a Late Classic temple at Calakmul, Mexico (Carrasco et al. 2009).

Traders

A sample of 15 Warrie sherds matched distinct geological locations in southern Belize suggesting multiple consumers (McKillop et al. 2018). Identifying the origins of the minor quantities of nonbriquetage pottery using petrographic analysis indicates that pottery was obtained from multiple areas, underscoring the likelihood that the Paynes Creek salt makers or local sea traders took their salt and other goods by boat to markets at various inland communities. The Ixtápa salt was sold by merchants from Zinacantán (Andrews 1983), whereas the salt makers at Sacapulas sold their own salt, visiting different towns (Reina and Monaghan 1981). The Paynes Creek salt workers may have sold their salt, salted fish, cacao, and marine resources. Hardened salt was traded in pots of standard size and dimensions in the Philippines, where potters made salt pots for specific village (Yankowski 2010). The size of the salt pots was standardized as a unit of trade for market. At San Mateo Ixtatán, salt was made in standardized pots that were broken to reveal the salt cakes as a standard exchange unit (Andrews 1983). The Paynes Creek salt maker/traders bought nice pottery and other supplies for salt production, corn and other food, and figurine whistles and other trinkets. Petrographic analysis indicates the briquetage was made from local clay and sand (samples found by the lagoon). Nonbriquetage pottery contained materials from at least 15 geological locations in southern Belize. Petrographic analyses may indicate marketplaces to which the salt makers or affiliated traders transported salt, salted fish, and other marine products.

Scale of Production

If all the salt works were in use at the same time, the scale of production would have been large, approaching industrial-scale production, although it may have been carried out by households. The scale of production would have been much smaller if only a few salt works were in operation at a time over the course of several hundred years. The first step in addressing the scale of production is estimating the dates of the individual salt works.

The settlement pattern at the Paynes Creek Salt Works consists of salt works along lines in the East and West Lagoons as well as a few larger sites with multiple wooden buildings and plazas, with Sites 7 and 60 being the largest. The location of salt works along shorelines may have provided ready access to salty lagoon waters. However, sea-level rise over the course of the Classic period may have encroached on the shoreline salt works, resulting in their abandonment and the construction of new salt works farther back from the lagoon on dry land. Since gradual sea-level rise has been documented during the Classic period in the area (McKillop 2002; McKillop et al. 2010a, 2010b), shoreline salt works subject to flooding from sea-level rise may have been abandoned and rebuilt farther back from the lagoon waters several times over the course of several hundred years. If this were the case, wooden buildings located in a line would date to the same time and would date to a different time from structures along an adjacent parallel line. Furthermore, the placement of palmetto palm posts may reflect attempts to maintain land from encroaching seas.

If rising seas and shoreline erosion were not responsible for the parallel lines of salt works, some buildings may have been residences located behind the salt works, as at Sacapulas and central Mexico (Parsons 2001), where salt workers' houses are located away from the salty workshops. In either scenario, "infilling" of salt works along shorelines may have occurred, with a few salt works initially located on a shoreline and later salt works located nearby, which would result in sites on the same shoreline dating to different time periods. If "infilling" occurred, I would expect later sites to fill in spaces between existing salt works. Lines of sites may reflect their placement on the edges of water bodies extant at the same time. There may have been a massive construction of salt works during the Late and Terminal Classic, using all available land.

Radiocarbon Chronology of the Paynes Creek Salt Works

Criteria used to select wood samples for radiocarbon dating were

to select one sample per site to maximize the spatial coverage across the lagoon;
to include samples from the East and West Lagoons;
to include samples from sites with mapped wooden architecture in preference to sites where wooden architecture was identified as present but not yet mapped; and
to investigate whether lines of wooden structures date to the same time and are different in age from adjacent lines of sites.

The radiocarbon dates on wooden posts include age ranges from the Early Classic to the Early Postclassic periods (Figure 6.18, Table 6.2).[2] The radiocarbon dates document when the tree was cut for the building post, which may have been when the building was first constructed or when it was renovated. In addition, it is possible that some building posts may have been scavenged from abandoned salt work buildings for new construction elsewhere. Wooden posts from four sites have radiocarbon dates with 2-sigma ranges from the Early Classic (Tzakol 3) to the beginning of the Late Classic (Tepeu 1).[3] One site includes an age earlier in the Early Classic (Tzakol 2) in the radiocarbon range. A post from Site 54 includes the end of the Early Classic in the 2-sigma range (see Figure 4.9a). A wooden building at Site 46 dates somewhere during the early part of the Late Classic (Tepeu 1). Seven sites had buildings constructed during the Late Classic Tepeu 1 or 2 (Sites 12, 25, 46, 48, 49, 67, 72; see Figures 4.9a, 4.12a). Posts from eleven sites have radiocarbon ranges that predate the Terminal Classic. Fifteen of the radiocarbon-dated posts from the Paynes Creek Salt Works were from trees cut sometime during the Late to Terminal Classic periods. Fourteen sites include the Early Postclassic in their age ranges.

None of the salt works has a radiocarbon range limited to the Early Postclassic (AD 900–1200). The 2-sigma range for radiocarbon dates of 14 posts extends from the Classic into the Early Postclassic. Only K'ak' Naab' (Site 14) has a radiocarbon range limited to the Terminal Classic and Early Postclassic (see Figure 4.12a). Six sites have radiocarbon ranges that extend from the Late Classic Tepeu 2 to the Postclassic. Seven sites have posts whose radiocarbon range extends from the Late Classic Tepeu 1 into the Postclassic. Site 10 has a radiocarbon range that includes the

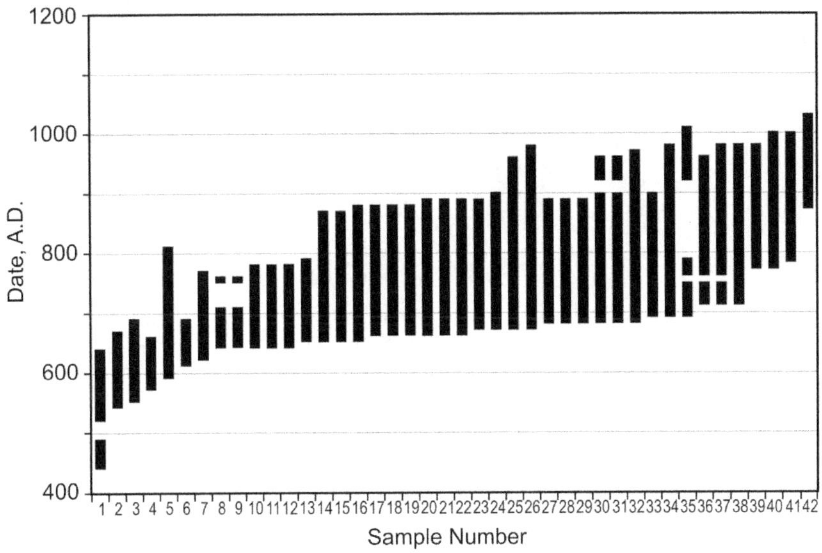

Figure 6.18. Radiocarbon dates from wood posts, 2-sigma, calibrated ranges.

Late Classic Tepeu 1 and 2 and the Early Postclassic but excludes the Terminal Classic (see Figure 4.12a).

Both the East and West Lagoons include radiocarbon-dated posts whose ages extend from the Early Classic to the Early Postclassic, indicating there was not a shift in settlement from one lagoon to the other over time (see Figures 4.9a, 4.12a). There are 20 radiocarbon-dated posts from the East Lagoon and 21 from the West Lagoon. However, there are some differences in dated posts between the two lagoons: There are more early dates in the West Lagoon and later dates in the East Lagoon. There are 17 posts from the West Lagoon predating the Terminal Classic, but only 10 posts from the East Lagoon predating the Terminal Classic. The median date for posts in the West Lagoon is from Site 75 (cal AD 650–880), whereas for the East Lagoon the median date is from a post from Site 11 (cal AD 670–960). The earliest radiocarbon date is from a post at Site 24 in the East Lagoon. Early Classic radiocarbon dates from posts include two sites in the East Lagoon (21, 24) and three sites in the West Lagoon (77, 70, and 54). Other C14 samples include the Early Classic in their 2-sigma ranges, notably, the canoe from Site 67, the wooden pottery-making paddle from Site 103, the mamey endocarp from Site 56, and oyster shells from Site 62.

Table 6.2. Radiocarbon dates, calibrated, 2-sigma, from Beta Analytic

Sample #	Site	Post #	C14 range AD, 2 Sigma, Calibrated
27	3	18	680–890
22	8	139	660–900
35	10	28	690–1010
25	11	191	670–960
10	12	3	640–780
24	13	2	670–900
42	14	Big post	870–1030
20	16	12	660–890
17	18	2	660–880
18	19	8	660–880
4	21	6	570–660
40	23	171	770–1000
1	24	50	440–490 and 520–640
7	25	74	620–770
37	26	1	710–750 and 760–980
30	29	4	680–900 and 920–960
34	30	1	690–980
38	32	5	710–750 and 760–980
28	34	1	680–890
21	35	2	660–890
6	46	44	610–690
19	47	63	660–880
8	48	36	640–710 and 750–760
9	49	5	640–710 and 750–760
23	52	12	670–890
5	54	13	590–810
31	56	145a	680–900 and 920–960
14	57	6	650–870
15	58	26	650–870
32	60	255	680–970
12	67	133	640–780
26	69	121	670–980
3	70	61a	550–690
11	72	67	640–780
39	73	2	770–980
36	74	72	710–750, 760–790, and 920–960
28	75	7	680–890
16	75	76	650–880
33	77	55	690–900
2	77	60	540–670

Table 6.3. Overlap in radiocarbon date ranges for lines of wooden structures, East Lagoon

Sites	Radiocarbon dates (calibrated, 2-sigma)
21, 24, 25	AD 570–640
23, 26, 19, 29, 32	AD 770–880
10, 34	AD 690–890
11, 22, 30[a]	AD 670–960
14, 13, 12, 16	AD 690–780

[a] Sites 22 and 30 are undated but in same line as Site 11.

Radiocarbon dating of wooden posts from salt works supports the hypothesis that salt works located along shorelines in the West Lagoon were abandoned due to sea-level rise and moved farther back from the lagoon water and rebuilt on dry land (see Figures 4.9a, 4.15). The radiocarbon dates for the wooden structures at Sites 46, 47, 48, 49, 72, and 75 along ancient shoreline 1 overlap in time between cal AD 660 and 690. The line of wooden structures at Sites 52, 54, 73, and 77 to the east on shoreline 3 have radiocarbon dates on posts that overlap in their ranges between cal AD 690 and 810. The ranges of dates for the two lines of sites do not overlap in time, supporting the hypothesis of resettlement due to rising seas, with the structures constructed near the water along shoreline 1 abandoned with rising seas, and new buildings constructed to the east along shoreline 3. There are insufficient radiocarbon dates for sites on other shorelines.

In the East Lagoon, there are four lines of wooden structures (see Figures 4.12a, 4.16; Table 6.3). A shoreline in the center of the East Lagoon formed by wooden structures at Sites 35/39, 36, 33, 21, 23, 25/26, 24, 19, 29, 32, and 31 do not have radiocarbon dates that all overlap in the 2-sigma ranges; Sites 21, 24, and 25 have ranges that overlap between AD 570 and 640. The other wooden structures have ranges that overlap between AD 770 and 880. Evidently some wooden structures were built first and other structures infilled at a later time. The wooden structures dated in the line of sites to the north along shorelines 3 and 4 and south along shoreline 1 overlap in their ranges with the dates for the central line of sites, along shoreline 2. There is only one radiocarbon-dated structure along shoreline 3 to the north at sites 22, 20, 15, and 11. Sites 10 and 34 along shoreline 1 to the south have radiocarbon dates that overlap in their ranges between AD 690 and 890. Farther north, wooden structures on shoreline 4 at Sites 13, 12, 15, and 16 have radiocarbon ranges on posts that overlap between

AD 690 and 780. Site 14 overlaps if the date for the K'ak' Naab' paddle is used instead of the date for a post. Except for the wooden structures at Sites 21, 24, and 25, all the sites, regardless of location, overlap in their radiocarbon date ranges. The data support an interpretation that wooden structures were initially constructed at Sites 21, 24, and 25 in the central part of the East Lagoon, with other wooden structures subsequently located on the same line and elsewhere. The spaces between the lines of wooden structures call for an explanation not revealed by radiocarbon dating. The spaces may have been waterways instead of the lines of sites developing sequentially as a response to moving location from rising seas.

Interpretations

Radiocarbon dating indicates that most of the wooden buildings at the Paynes Creek Salt Works were constructed during the Classic period (AD 300–900) when inland demand for salt was at its maximum. Radiocarbon ranges for posts suggest that there was initial construction of several salt works toward the end of the Early Classic or the beginning of the Late Classic. The date ranges suggest that construction began in the Early Classic in both lagoons but that there was more early construction in the West Lagoon and later construction in the East Lagoon. Radiocarbon dating supports an interpretation of moving salt works to drier land in response to rising seas in the case of two adjacent lines of wooden structures in the West Lagoon. However, the location of salt works of different ages along the same line of sites in the East Lagoon suggests "infilling" to use all available space close to the salt water.

Radiocarbon dates on wooden posts range from the end of the Early Classic (AD 300–600) through the Late Classic (AD 600–800) and Terminal Classic (AD 800–900) and into the beginning of the Postclassic period. All dates include the Late Classic and Terminal Classic in their 2-sigma range, so the radiocarbon dating does not address the issue of scale of production by specifying the age of each salt work. Of course, some salt works may be multicomponent as well. The 2-sigma radiocarbon dates include a range of about 200 years, which is too long a time interval for estimating the construction sequence of wooden buildings at the Paynes Creek Salt Works. Study of changing pottery styles was used to help refine the radiocarbon chronology.

Radiocarbon dating sediment columns indicates sea-level rise continued during the use of the salt works and afterward. Postclassic dates on marine sediment in the Postclassic record gradual sea-level rise that buried the salt works in red mangrove peat. After that time, sea levels rose more rapidly, flooding the red mangroves and creating the open lagoon.

Ceramic Chronology

The location of ceramic types, vessel forms, and vessel traits characteristic of specific times were mapped to identify temporal patterns in use of salt works within the lagoon system. Some salt works may have been used longer than others. The linear pattern of salt works suggests they were built on ancient shorelines that became flooded and resulted in relocating the salt works along a new shoreline.

Some temporal markers for pottery elsewhere in the Maya area may provide clues to chronology at the Paynes Creek Salt Works. Late Classic markers include the presence of vases, Lubaantun-style figurine whistles, and red-slipped serving bowls with basal angles. Late to Terminal Classic markers include Belize Red pottery, grooved or square lips on Mangrove Unslipped jars, and the presence of notched basal angles on Belize Red bowls. Terminal Classic markers include incised panels below the rim of vases and gouge-incised decoration. Ceramic markers indicate several sites have a Terminal Classic component, including a partial vase with incised glyph panel below rim from Site 7, a gouge-incised jar fragment from Site 61, and a gouged-incised vessel from Site 69 (Figure 6.3). A tripod bowl with trickle painted decoration was recovered from Site 56 (McKillop 2011:Figures 11–12). Some sites may date to the end of the Early Classic or beginning of the Late Classic, at least in part due to the presence of Dos Arroyos Orange pottery at Site 67 and Paradero Fluted pottery at Site 72.

Standardization of Salt-Making Vessels as a Measure of Mass Production of the Product, Salt

Salt was boiled in standardized containers for ease of use during evaporation and to produce a standard-size block of salt for market. Modern and historic accounts of brine boiling show many clay pots over a fire at

Sacapulas and Bohol Island in the Philippines. In Bohol, pots are made by potters in standardized sizes and shapes for the salt makers "so that the salt cakes are of standard size and weight, making them a fixed unit of trade" (Yankowski 2010:177). Andrea Yankowski (2010:177) writes, "The standardization of production methods and the form and size of the earthenware pots, which is aimed to create uniform units of trade, reflects the market oriented nature of this industry, in which salt was traded as a valued commodity." In the western salt-making towns, salt pots are 13 cm high, with a maximum body diameter of 10 cm. One pot of salt is valued at 7.8 kg of rice or 50 local pesos. In the eastern towns, pots are 17 cm high and 15.5 cm in maximum body diameter. One pot of salt is valued at 46.8 kg of rice or 300 local pesos (Yankowski 2010:173–175).

At Santa Filomena, Bohol, Santa Fe, and Albuquerque in the Philippines, salt is made by soaking chopped coconut husks and palm tree trunks in salty water, drying it, and then leaching the brine (Yankowski 2010:Color Plates 8.1, 9.1–9.2, Figure 2). In Albuquerque, the salt is hardened in small jars similar to those at the Paynes Creek Salt Works and then traded that way, sometimes with sherds removed from the base to reveal the salt. In northeast Thailand, the size of stoneware jars and baskets are standardized for selling salt, which is often traded in baskets (Yankowski and Kerdsap 2013:241).

At San Mateo Ixtatán, in the Huehuetenango department in the highlands of Guatemala, brine boiling was done in small, open vessels that cracked and produced a solid 22 kg salt cake (Andrews 1983:Figures 4.9–4.10). At Ixtápa in the Maya highlands of Chiapas, Mexico, near Zinacantán and San Cristobal, brine boiling was formerly done in large clay pots that were about 70 cm in diameter and about 90 cm deep. When the salt began to congeal, it was packed into straw tubs to produce a standardized salt loaf of about 600 g (Andrews 1983:58–59; Figure 3.6). Ixtápa dates as early as the Late Preclassic.

Standardization of briquetage in a previous study of the first three sites excavated at the Paynes Creek Salt Works (McKillop 2002:127–133) suggests mass production of salt based on standard containers. Other indices, including richness (number of pottery types) and evenness (number of sherds per type), indicate that the sites were focused on salt production and were not residences: Sites 56, 57, and 58 had few pottery types, and most of the pottery was Punta Ycacos Unslipped pottery used in salt production (McKillop 2002:Table 3.10). Similar results were found for pottery at two additional salt sites, Chan b'i (Site 24) and Atz'aam Na (Site 35) in the East Lagoon (McKillop and Sills 2017:Table 4.1; Sills and McKillop 2013, 2018a). The presence and abundance of briquetage versus other pottery and artifacts was used in southeast China to distinguish residences from salt kitchens (Flad 2013).

Study of the briquetage from three salt works discovered and excavated in a previous study (McKillop 2002) indicates that the salt pots were locally made by the Paynes Creek salt workers as needed at individual salt works: Evidence is the friable nature of the salt pots, the availability of local clay and sand used to make the salt pots, and the statistically significant differences in standardization among three of the salt works—namely, Stingray Lagoon, David Westby, and Orlando's (McKillop 2002:127–133, Table 4.1 4.6). In the previous study, using the average median variation, the diameter of Punta Ycacos jar and bowl openings and the diameter of solid clay cylinder vessel supports were found to be standardized in comparison with a sample of Bedford Unslipped vessels from household deposits from nearby Wild Cane Cay. The differences were statistically significant. In fact, there was twice as much variation in the Wild Cane Cay vessels in contrast to the Punta Ycacos bowls, jars, and cylinders. Furthermore, there were statistically significant differences in standardization among the salt kitchens at Stingray Lagoon, David Westby, and Orlando's. These differences were interpreted as separate salt works used by different work parties, or perhaps households.[4]

Spatial variability in the shapes of salt pots among different workshops as well as differences in standardization of the vessels among sites may indicate that different work parties or households used specific salt works. Since the briquetage was locally made from local clays and quartz temper, specific pottery vessel shapes may have been made at some sites for their use or for exchange. Five areas of the lagoon system contain most of the salt works. These areas are useful for spatial comparisons.

The sample of briquetage for assessing standardization consists of pottery collected before and after the wooden buildings were discovered, with different field objectives. The goals of the pedestrian survey in 2003 and 2004, before the discovery of the wood architecture, were to see whether there were more salt works than the four previously excavated (McKillop 2002) and, if there were more sites, to locate about 30 measurable rims and 30 measurable vessel supports made from Punta Ycacos in order to evaluate variability in salt production in comparison with the previous

Table 6.4. Diameter measurements of clay cylinder vessel supports from the 2003–2004 sea floor survey

Site	Year	# Items	Mean	Standard deviation	Coefficient of variation	Median
66[a]	2003	25	3.484	0.14	14.14	—
61	2003	21	3.967	0.28	7.13	—
65	2003	20	3.64	—	—	—
64	2003	24	3.942	0.28	7.18	—
62	2003	43	3.84	0.42	11.05	—
60	2003	57	3.97	0.56	14.25	—
67	2003	51	4.3	1.5	36.18	—
1	2004	11	4.7	0.07	1.5	3.9
2	2004	11	5	0.64	12.8	5.15
3	2004	6	4.3	0.78	18.1	4.4
4	2004	7	5	0.49	9.8	4.6
5	2004	9	4	0.28	7	4.4
6	2004	15	4.3	1.2	27.9	4.4
7	2004	8	8	—	—	—
8	2004	15	4.5	0.49	10.9	4.3
9	2004	10	5.2	0.71	13.7	5
10	2004	11	4	0.5	12.5	3.8
11	2004	44	4	0.14	3.5	3.9
12	2004	23	4	0.21	5.25	4
13	2004	34	3.9	0.57	14.6	3.5
14	2004	12	3.9	0.07	1.8	4.1
15	2004	23	3.8	0.14	3.7	3.6
18	2004	28	4	0.28	7	4.1
19	2004	28	3.9	0.07	1.8	4
20	2004	9	4	0.21	5.25	3.9
21	2004	5	4.1	0.64	15.9	4

[a] Site numbers assigned starting at 1 in 2004, with sites found previously assigned numbers afterward.

sites. Once a site was located, artifacts were collected, a PVC site marker was embedded in the sea floor near the center of the artifact distribution on the sea floor, a water depth was taken, and the team continued the search. The artifacts were located to site only, were drawn, and artifact observations and measurements were recorded. Each artifact was assigned a number when it was drawn. The numbered artifacts were subsequently added to the GIS with a notation that they were not individually mapped.

Table 6.5. Briquetage rim sherd diameters from 2003 sea floor survey

Site	Shape	Year	# Items	Mean	Standard deviation	Coefficient of variation	Median	Thick cm	Mean % diameter
60	bowl	2003	9	30.1	12.73	42.3	33	1.3	24.4
67	bowl	2003	18	25.1	2.1	8.4	15.5	0.9	17.1
62	bowl	2003	3	32			32	1.2	6.7
61	bowl	2003	5					0.8	
65	bowl	2003	1	28			28	1.3	10
60	jar	2003	22	22	9.9	45	19	1	20.7
67	jar	2003	13	23.3	2.8	12.1	18	1	8.7
62	jar	2003	4	22.3			22		11.3
61	jar	2003	8	20.3			20	1.2	18.4
65	jar	2003	1	14			14	0.7	16
63	jar	2003	4	22			19.5	1.7	12.3
64	jar	2003	4	20			21	1.7	9.3
66	jar	2003	2	27			27	2	7

The numbered artifacts provide a sample of Punta Ycacos rims and clay cylinder vessel supports for evaluating standardization of production. Previously, I suggested there were different work parties, perhaps family or extended family groups, based on differences in the average median variation among three sites (McKillop 2002). The numbered artifacts provide a comparison. The 2003 and 2004 survey data indicate there was variability in the standardization of the clay cylinder vessel supports at the 26 sites, including sites from the East Lagoon, Main Channel, and West Lagoon (Table 6.4).

The coefficient of variation shows the pots were not made at one location and distributed (Table 6.5). Instead, pots were made by salt workers for their individual salt works. This parallels the practice in Sacapulas, where pots were made at home in advance of evaporation. At Sacapulas, only the open bowls used for evaporation that were broken to remove the hardened salt cakes were noted as made at home by Ruben Reina and John Monaghan (1981). The high value of the coefficient of variation for Site 7 cylinders and Site 60 pots may be due to the sample size but also to the fact that Site 7 and 60 are larger sites with multiple buildings and plazas. Some of the buildings at Sites 7 and 60 may have been used for residential or other uses, as previously discussed.

Tree Species Selection, Standardization of Briquetage, and Radiocarbon Dates on Posts

The selection of tree species for building construction points to work parties who selected and constructed salt production "kitchens." Standardization of briquetage indicates that the salt workers at each salt kitchen made their own pottery. Tree species identification of wood posts from the three sites, combined with radiocarbon dates, lends further support to the interpretation that the salt works were used by separate work parties (Figure 6.19). In fact, tree species identifications for posts and radiocarbon dates suggest that the David Westby and Orlando's sites were constructed at the same time, whereas Stingray Lagoon was later and also likely represents two separate construction efforts and perhaps distinct "sites." The Stingray Lagoon site includes two clusters of posts and associated artifacts, including a northern area and a southern area. The northern and southern areas of Stingray Lagoon are separated by more than 10 m, which normally would have meant they were designated as separate sites, but the space was not noticed during field research. Both David Westby and Orlando's salt works have Santa Maria, rosewood, *Mouriri* sp., *Casearia* sp., and white mangrove. In contrast, Stingray Lagoon North has *Mouriri* sp., Chrysobalanaceae, and *Casearia* sp., and Stingray Lagoon South is dominated by Santa Maria.

Stingray Lagoon is 300 m distant from the other 2 sites, so it is not surprising that it is distinctive. Stingray Lagoon has a Terminal Classic tripod bowl with trickle-ware paint dripped on the exterior (McKillop 2011:Figures 11–12). A Belize Red bowl with exterior scallops on the panel above the basal break is the only one of its kind at the Paynes Creek Salt Works. There are Lubaantun-style figurine whistles, a large figurine whistle depicting a woman (see book cover), and a bat whistle from previous excavations (McKillop 2002:Figure 3.40). Previous excavations also yielded a Warrie Red jar rim with monkey stamps and comb stamps around the vessel shoulder and lower on the vessel body (McKillop 2002:Figure 3.27).

Three salt works along a line, perhaps an ancient shoreline, in the East Lagoon were selected to examine patterns of tree species selection for building construction along with radiocarbon dates. The results indicate that Site 23 is later than Sites 24 and 25/26, which may be contemporaneous. Sites 24 and 25/26 are dominated by posts made from black mangrove

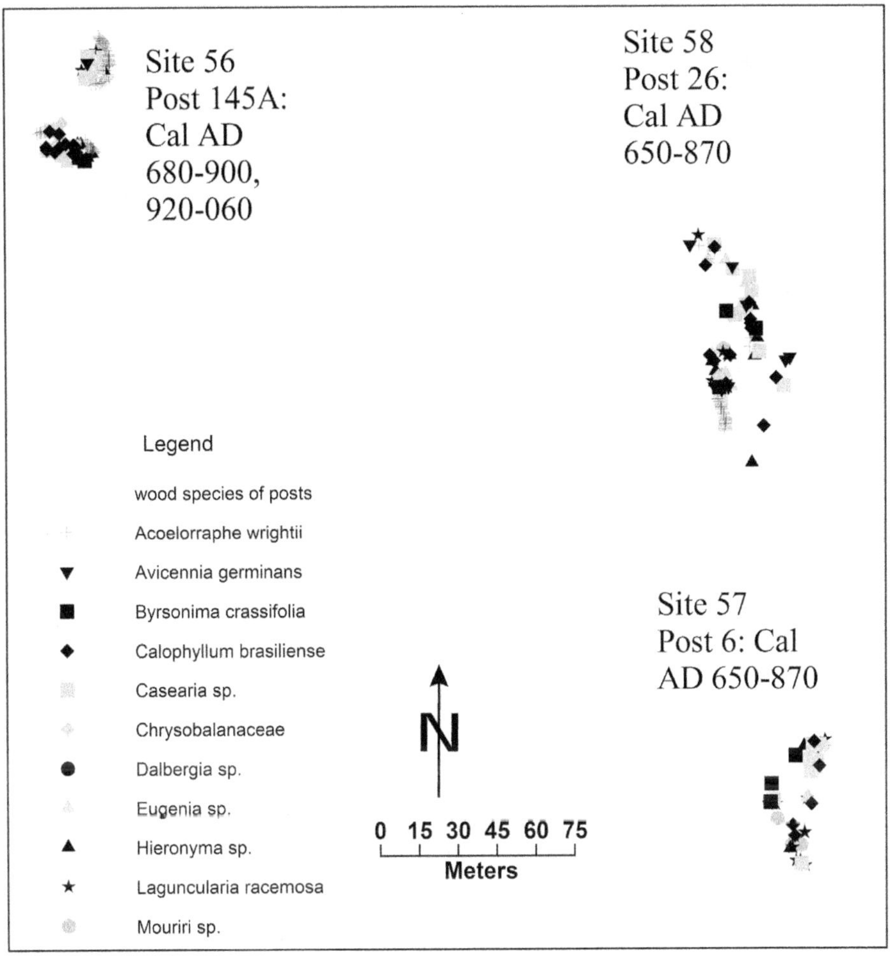

Figure 6.19. Wood species selection for salt kitchens and radiocarbon dates in the Stingray Lagoon area.

and white mangrove, whereas Site 23 is dominated by Santa Maria, with some rosewood, but lacks black mangrove (Figure 6.20).

Farther east, Sites 31 and 32 share *Hieronyma* sp., Chrysobalanaceae, and Santa Maria. *Mouriri* and *Casearia* are at Site 31; White mangrove is at Site 32. To the west at Site 35, *Mouriri*, *Eugenia*, Santa Maria, and *Casearia* are present. Site 60 posts are mainly *Mouriri*, Santa Maria, and Chrysobalanaceae, with white mangrove, craboo (nance), *Eugenia*, *Casearia*, and rosewood also present.

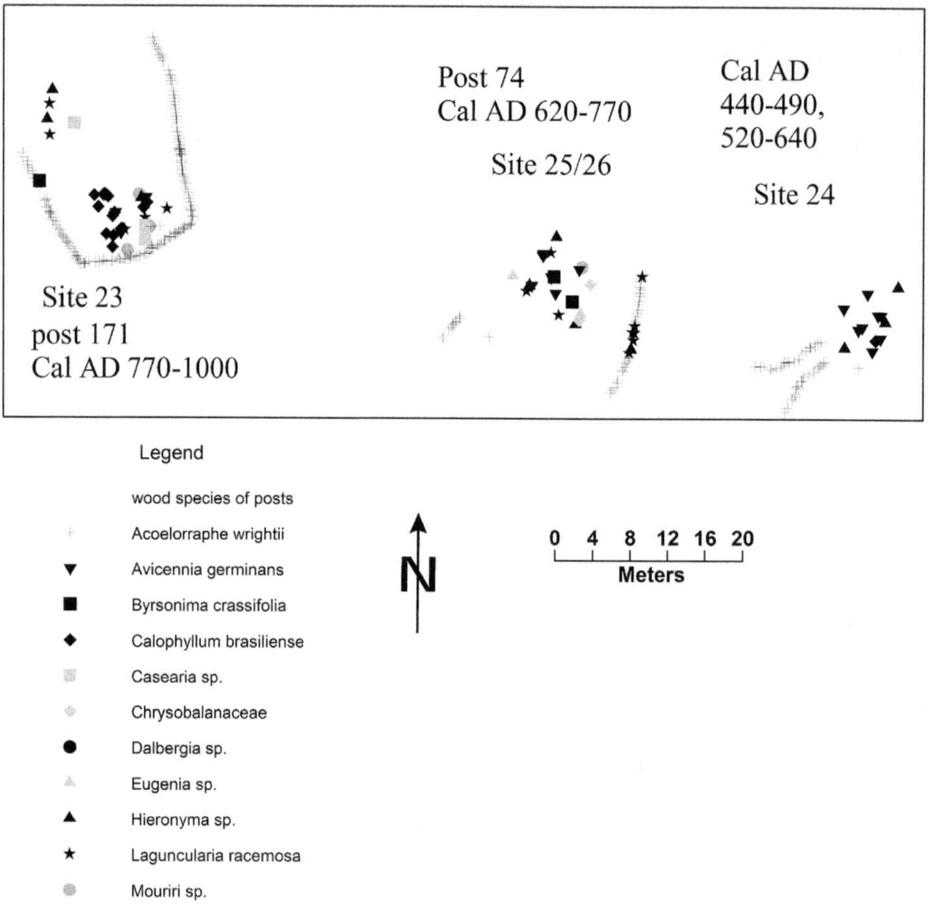

Figure 6.20. Wood species selection for salt kitchens and radiocarbon dates for three adjacent sites on a line in the East Lagoon.

The West Lagoon is distinct from the East Lagoon and Stingray in tree species selection for building posts. In the West Lagoon, Santa Maria, rosewood, and *Mouriri* are dominant. Other species include Chrysobalanaceae, *Casearia*, *Eugenia*, *Hieronyma*, and craboo. White mangrove is present in small numbers. Striking differences within the West Lagoon include the high number of rosewood posts at Site 72 and the high number of Santa Maria at Site 47 and 75, separated by several sites. Rosewood and Santa Maria also are present at Site 74, adjacent to Site 72. Site 77, farther east, lacks rosewood or *Mouriri* but shares Santa Maria, craboo, *Hieronyma*, and Chrysobalanaceae with Site 74.

Estimating the Amount of Salt Produced

Salt production estimates at the Paynes Creek Salt Works were calculated using the Sacapulas salt works as an analogy. Reina and Monaghan (1981) observed that two people made 250 pounds (113.4 kg) of salt in two days in a 6 × 8 m salt "kitchen"—a building dedicated to salt making—at Sacapulas. This estimate includes the time to haul baskets of salty soil, leach the salt by pouring water through the salty soil, evaporate the enriched brine in 24 open bowls over a fire, and further harden the loose, wet salt into two types of salt cakes. Salt production at Sacapulas was a full-time, seasonal activity carried out in the dry season because the salt playas were flooded in the rainy season. However, the salt workers piled salty soil in the salt kitchens so salt could be made whenever it was needed. During their 1981 study, Reina and Monaghan (1981:17) noted that 33 people were engaged full time making salt at 11 kitchens. Before a rock slide in 1932 that reduced the area of the salt playas, there were 50 full-time salt workers working more kitchens.

At Sacapulas, two to three people could make 750 pounds (340.2 kg) of salt in a six-day week or 12,000 pounds (5,443.2 kg or about 6 tons) of salt in a four-month dry season. Applying this estimate to the Paynes Creek Salt Works, if 10 salt works were used by 20–30 people at a time, salt yields would be 7,500 pounds (3,402 kg) per week or 120,000 pounds (54,432 kg or about 60 tons) over a 16-week dry season.

The demand for salt by the ancient Maya included dietary salt, salt for drying fish and other meat, and salt used as a flavor enhancer. The carbohydrate-rich diet of corn, beans, squash, and other cultigens and tree crops did not provide dietary salt. Some salt was certainly obtained from eating animal meat, but that formed a small part of the diet. Salt obtained from burning palm trees reduced the supply of palm fronds for thatching roofs and was arguably not a major source of salt (Marcus 1991; McKillop 1996b, 2002). That the Classic Maya also used salt to dry fish is now documented by use-wear analysis of chert stone tools at the Paynes Creek Salt Works (McKillop and Aoyama 2018). Salt-drying fish may have occurred inside some of the wooden buildings, to keep the fish dry and to promote drying from smoke and heat produced by the salt evaporation. Fish-drying racks may also have been placed in the yards.

Since salt is an addictive condiment, a flavor enhancer, and one of the four taste sensations on the human tongue, the Maya may have craved salt beyond its biological needs. The amount of dietary salt needed depends on a person's level of physical activity, humidity, and body mass. Among various estimates of daily salt needs (e.g., Andrews 1983; Vogel 2010), Samuel Adshead's (1992:141) estimate of 6 g per day per person is used in this study to evaluate salt production and consumer needs. The data may be varied by others using other estimates of salt needs, such as Anthony Andrews's (1983:110) estimate of 8 g per day per person for the Classic Maya.

Using 5 pounds (2.3 kg) per capita per year estimate of biological dietary salt needs (6 g/day; Adshead 1992:141), the Paynes Creek Salt Works produced salt for 24,000 people, with 10 salt works in operation for four months per year. If 20 salt works were in operation by 40 to 60 people, salt production would satisfy the dietary salt needs of 48,000 people.

If some of the salt was used for salt-drying fish, as indicated by use-wear analysis of chert tools, production estimates for dietary salt production would have been lower: If 20 salt works were in use at a time, and half the salt production was used for salt-drying fish, that would still leave 120,000 pounds (54,431 kg or about 60 tons) for dietary salt production.

At the height of Late and Terminal Classic period, when population at nearby inland Maya communities such as Lubaantun, Nim Li Punit, and Pusilha was greatest, perhaps up to 100 of the Paynes Creek Salt Works were in operation, yielding 75,000 pounds (34,019 kg) of salt/week or 1,200,000 pounds (544,310.8 kg or about 600 tons) over four months. That amount would have provided dietary salt for 240,000 people, or less if some salt was used for preserving fish. Nonetheless, salt production exceeded nearby Maya dietary salt needs (McKillop and Aoyama 2018).

Salt allows the accumulation of surplus food (salt cakes, salt-dried fish/meat, and fish sauce) for local use and trade, including long-distance trade (Yankowski et al. 2015:9). Salt facilitates dietary flexibility to compensate for the seasonality of local foods by storing salted fish or by storing salt cakes for trade of inland corn during food shortages, making surplus salt production a good risk-management strategy.

Ocarinas

The identity of the salt workers may have left markers in the material record. The three salt works in the Stingray Lagoon area have statistically significant differences in the standardization of the salt pots and cylinders,

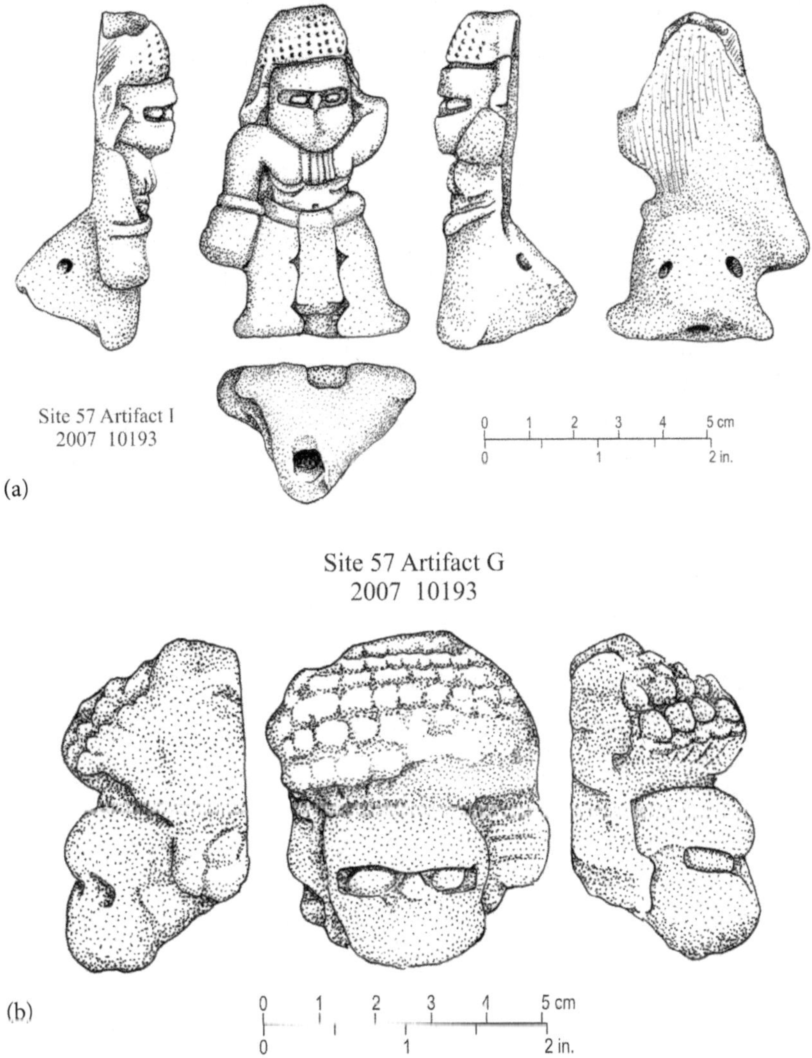

Figure 6.21. Helmeted figure with distinctive "grape bunch" headdress from Site 57I (*a*) and head only from Site 57G (*b*). By Mary Lee Eggart.

suggesting that different work parties were responsible for making the salt pottery and producing salt (McKillop 2002:133). Iconographic differences in the ocarinas also points to different identities: Site 57 has two examples of a different ball player with a distinctive headdress, perhaps identifying the Site 57 salt producers who took their salt to market and traded independently for goods (Figure 6.21). This style of headdress is not found among the helmeted figurine whistles at Site 60, where a variety of other headdress styles is found (Figure 6.4). Trade goods, especially figurine whistles, may point to the identity of different salt production units.

Summary

Salt production at the Paynes Creek Salt Works fits a model of surplus household production, with a variety of other related activities that can be described under the rubric of contingent of multicrafting. In addition to enriching the salinity of the brine by pouring water over salty soil (which had to be obtained), the salt workers boiled the brine in nearby salt kitchens. Contingent multicrafting activities for which there is evidence at the sites include making the salt pots, obtaining wood and constructing the salt kitchens, obtaining firewood, boat transportation, eating, cutting fish and/or meat and hides, and probably salting them for trade, and taking salt cakes, fish, salted hides, and other marine resources by canoe up rivers to inland marketplaces. Year-round use of the salt works is suggested by the recovery of the remains of tree fruits that ripen throughout the year, by the significant investment in buildings and equipment to maintain, and by the year-round market for dietary salt and salted fish.

The Paynes Creek Salt Works provided dietary salt to meet the needs of the nearby inland Maya at cities and other communities inland in southern Belize as well as to produce salted fish. Both were storable commodities. In addition, salt cakes produced as standard units of exchange retained their value as exchange equivalencies in subsequent transactions. Applying the estimates to other salt works in the Maya area indicates that the regional trade of salt from coastal and inland salt works provided sufficient dietary salt for the Maya at nearby communities and, in the case of the coastal salt works, provided salted fish and other marine resources for marketplace trade.

7

Marketing Salt, Salted Fish, Cacao, and Marine Products

What were the mechanisms for inland supply of salt from the Paynes Creek Salt Works? In ancient China, the emperor sometimes placed his representatives in local industries distant from the capital in order to directly manage production and supply of salt (Adshead 1992; Flad 2007; Vogel 2010). The Aztecs annexed new territories and extracted tribute, with chocolate an exception; sometimes chocolate was obtained by trade with the Soconusco region instead of by tribute, perhaps due to its remote location and therefore difficulty in maintaining control from Tenochtitlan. The Inca moved artisans to their capital at Cuzco and exacted tribute from provincial craft workers. Was salt produced along the coast of Belize and transported to inland cities as tribute, as taxation, or as commodities for the marketplace?

Surplus household production at the Paynes Creek Salt Works fits with the transport of salt cakes, salted fish, and maritime resources up rivers to marketplaces. Salt producers may have taken a load of salt to a marketplace on a regular basis, knowing the periodicity of markets in various towns (Freidel 1981; Freidel et al. 2016; Smith 1974). Nearby inland communities include Nim Li Punit and Lubaantun as well as coastal communities and inland and distant locations farther away (see Figure 1.2, Table 7.1). Salted fish, cacao, stingray spines, shellfish, and other resources produced or acquired on the coast may have been transported separately by different producers or in one boatload. There were trade goods of various kinds in the trading canoe encountered by Christopher Columbus off the north coast of Honduras near the Bay Islands on his fourth voyage to the New World.

Table 7.1. Distance from Paynes Creek Salt Works (Site 60 in the Main Channel) by water travel to other locations

Water travel distance to	Distance km
Wild Cane Cay	8
Frenchman's Cay	14.5
Green Vine Snake	16
Village Farm	20
Pork and Doughboy Point	22
Lubaantun via the Rio Grande	67
Rio Motagua	70

Wooden posts, thatch, clay, pottery temper, food, and drinking water were transported to the salt works by dugout canoes. There are natural springs or seeps in the immediate area, in addition to two large freshwater ponds north of the salt works and up rivers (see Figure 1.2). Many daily activities, such as bathing and washing clothes and dishes, do not require fresh water and can be accomplished with salt water. In addition, islanders dig holes to the freshwater lens that rests on the salt water. The freshwater lens was encountered in stratigraphic excavations at Wild Cane Cay, with salt water below (McKillop 2005b).

Cacao may have been planted by the coastal Maya in the Port Honduras mainland for local use at Frenchman's Cay and elsewhere as well as for transport up rivers to inland markets. In southern Belize cacao is documented from a Classic period midden under a coral building foundation at Frenchman's Cay (McKillop 2005a) and inland dated to about AD 300 at Bats'ub Cave, described as a mortuary cave 25 km north of Uxbenka (Prufer and Hurst 2007). The suitability of the Port Honduras coastal soils for cacao is attested by the presence of an organic farm at Village Farm with some 5,000 cacao trees producing for market. Certainly, cacao also is planted for market production inland in southern Belize. However, boatloads of cacao in sacks would have been easier for farmers to bring to markets than portaging by overland porters. Coastal traders may have had an advantage in selling their cacao.

Water distances can be traveled with ease by mariners: A group of women from Punta Gorda, Belize, paddled across to Puerto Barrios, Guatemala, and back in a single day (Iris Vernon, personal communication, 2016). That's generally a 2½ hour ride in a 35-foot skiff with two

200-horsepower engines by the ferry. The barrier reef is 40 km offshore from Punta Gorda, but one person in a small paddling canoe can make the trip, even in rough water at night, as local fisher Orlando Usher often accomplishes. Fisherfolk paddle small canoes offshore to fish for sale in the market in Punta Gorda. Water transport opens opportunities for travel, including salt, cacao, corn, and other bulk commodities for those who know the sea (McKillop 2007a, 2010b, 2016b, 2017b; McKillop et al. 2014). Ballast is important for traveling by canoe in order to keep the boat from drifting sideways, underscoring the likelihood that salt traders moved salt and marine resources to markets and returned with boats laden with other commodities from markets.

The Economics of Transport of Maya Commodities

There are several models of political economies elsewhere that might explain the organization of the Paynes Creek salt industry. The models include the tribute model, the alliance model, and the surplus household production model (McKillop 2009). The tribute model parallels the Aztec or Inca strategy of using military force or imposing local rulers. Following the tribute model, the inland dynastic Maya controlled the coast, via the trading port of Wild Cane Cay, and its resources, such as the Paynes Creek Salt Works, by incorporating the coast into a regional state where hinterland communities owed tribute. In the alliance model, the inland dynastic Maya may have created trading and other alliances, sanctified by rituals and feasts, in order to maintain a regular supply by trade in salt and access to external sea trade routes. Some areas elsewhere in Mesoamerica fluctuated between alliance and tribute, such as the Pacific coast Soconusco area during the Aztec empire (Voorhies 1989). In the surplus household production model, households produced a surplus of one or more types of goods. Surplus production may have been exchanged among households producing different goods, with larger-scale surplus production for trade at markets (Smith 1974) and fairs (Freidel 1981) nearby or farther away.

The Late and Terminal Classic Maya polities of southern Belize, closest to the Paynes Creek Salt Works, were decentralized, putting the coastal Maya in an advantageous position both economically and politically (McKillop 2005b, 2009). The inland communities with dynastic records—notably, Nim Li Punit, Lubaantun, Uxbenka, and Pusilha—were modest in size and did not reference one another in dynastic records on carved monuments (Braswell et al. 2004, 2007; Prager and Braswell 2016; Prufer et al. 2007, 2011). Local clays were used for making most pottery with intracommunity rather than intercommunity exchange (Jordan and Prufer 2017), underscoring the lack of centralized power in southern Belize during the Classic period.

If the dynastic Maya exacted tribute from subsidiary settlements within their realm and from conquered lands beyond, then food and salt may have been included as tribute items. Warehouses would have been needed to store bulk food. Pictorial depictions of royal visits and feasts on painted pottery vessels and murals and carved on building facades include elite goods produced at royal courts for exchange with dynastic and other noble Maya as well as tribute or trade items, including cacao and perhaps salt. The ancient Inca exacted a variety of kinds of tribute and placed warehouses throughout their empire to store locally produced food to feed the military and other state workers (DeLeonardis 2011). If the inland dynastic Maya exacted salt as tribute from the Paynes Creek Salt Works, then there would be evidence of some mechanism to enforce tribute payment, such as a dynastic representative to oversee production and taxation, as in ancient China. Enforcement would have included at least the threat of military intervention. The residence of a dynastic representative would be distinctive from other residences or salt works and might include an artifact assemblage with nonlocal and elite goods. No such evidence has been found at the Paynes Creek Salt Works.

Instead, evidence points to surplus household production of salt that was transported to markets elsewhere from independent salt kitchens at the Paynes Creek Salt Works. The diversity of inland origins of pastes on unit-stamped pottery found at the Paynes Creek Salt Works points to separate canoe trips to markets in various inland communities to sell salt at marketplaces (McKillop et al. 2018). The fact that Nim Li Punit, Uxbenka, and Pusilha had hieroglyphic records of their dynastic leaders on carved monuments but did not mention other nearby sites with dynastic records underscores the decentralized geopolitical landscape of inland southern Belize, from the Early Classic center of Uxbenka through the Late and Terminal Classic at Nim Li Punit and Pusilha as well as Lubaantun, which has glyphs on ball court markers but lacks stelae (Braswell et al. 2004, 2007; Prager and Braswell 2016; Prufer 2005; Prufer et al. 2007, 2011). Jillian Jordan and Keith Prufer (2017) describe utilitarian pottery production at Uxbenka, Nim Li Punit, and Lubaantun as self-sufficient within each center, based on household specialization of vessel form. Two settlement groups on the periphery of Uxbenka produced Remate Red (similar to Warrie Red) and Turneffe Unslipped (similar to Punta Ycacos Unslipped) pottery, respectively (Jordan and Prufer 2017:77–78). These households were located on soils that were clayey and thus good for pottery production but not for agriculture. Householders at Uxbenka, Nim Li Punit, and Lubaantun each produced their own utilitarian pottery, which was distributed within each site but evidently not among the three communities (Jordan and Prufer 2017:81). Compositional analysis of utilitarian and polychrome pottery from Norman Hammond's project at Lubaantun indicates the pottery was made with local raw materials (Hammond et al. 1976). Specialized household production of pottery by households and intracommunity distribution, together with different patterns of obsidian source use (Daniels and Braswell 2014), points to a lack of economic interaction between Nim Li Punit and Lubaantun (Jordan and Prufer 2017:81–82).

To Market by Maya Canoes

Apart from discoveries at the Paynes Creek Salt Works, there is no other evidence of actual ancient Maya canoes or paddles (McKillop 2010b, 2017b; McKillop et al. 2014). The Paynes Creek evidence of canoe travel includes the K'ak' Naab' canoe paddle, another paddle from Site 7, parts of canoe-paddle blades from the Elon site (Site 83) and Site 74, and the canoe from the Eleanor Betty site (McKillop et al. 2014). The discovery of a wooden canoe paddle from the K'ak' Naab' salt work provided the first tangible evidence of ancient Maya waterborne navigation. The discovery of the canoe from the Eleanor Betty site is the first report of an actual Maya canoe (McKillop et al. 2014). The canoe and paddles tie the production of salt to its transport in canoes. In addition to 4,042 wooden posts, beams, and wedges with worked ends, wooden objects recovered from survey indicate woodworking was part of the salt industry. The use-wear evidence that some of the chert tools were used in cutting and whittling wood indicates boat construction and canoe-paddle production at the Paynes Creek Salt Works.

These discoveries underscore the importance of canoe trade and travel during the Classic period in the Maya area. There was canoe travel within coastal areas such as Port Honduras and Paynes Creek Salt Works area, along the coast, and between the coast of Belize and inland Classic Maya communities. Sixteenth-century Spanish explorers to Central America disrupted ancient Maya canoe trade and travel that arguably had endured for more than a millennium. This included sea trade around the Yucatán but also shorter distance travel along inland waterways in Belize, Mexico, and Guatemala supplying the inland Maya at large urban cities with coastal resources (Andrews et al. 1989; Graham 1989; Graham and Pendergast 1989; Hammond 1972; Healy et al. 1984; McKillop 1996a, 2004; McKillop et al. 1988). Stingray spines for ritual bloodletting, seafood, shells for ornament and food, and quantities of salt were transported inland (McKillop 2007a). The Paynes Creek Maya were engaged in production and distribution of quantities of salt on the coast of southern Belize for inland trade up nearby rivers to Late Classic inland communities in Belize.

Prior to the discovery of the K'ak' Naab' paddle, our knowledge of canoe travel was based on artistic depictions of canoes and paddles (Hammond 1981; Marcus 1983; Thompson 1951; Trik 1963), settlement of offshore islands, and models and hypotheses about the timing and importance of ancient Maya sea trade (Andrews et al. 1989; Graham and Pendergast 1989; Guderjan and Garber 1995; Hammond 1972; Healy et al. 1984; McKillop 2004, 2016b; McKillop et al. 2004; McKillop and Winemiller 2004; Sabloff and Rathje 1975). Settlement of offshore islands underscores ancient Maya familiarity with boats and sea travel. Boats were necessary for transportation to island sites located off the coasts of Belize and the Yucatán including near-shore islands such as Isla Cerritos, Wild Cane Cay, Moho Cay, and Frenchman's Cay; islands located farther from the mainland such as Ambergris Caye and Cozumel; sites over 40 km offshore on the Belize barrier reef such as Hunting Cay; and sites on atolls beyond the reef.

Artistic depictions show paddles and their use in canoes, with most images showing sacred activities, such as the ancient Maya paddler gods' role in delivering the maize god from the primordial sea at the moment of creation of the world. Pictorial depictions incised on bones from Late Classic burial 116 in Temple 1 at Tikal (Trik 1963) show the Stingray Paddler god and the Jaguar Paddler god paddling a canoe (Figure 7.1). Like the K'ak' Naab' paddle, their paddles have a straight handle without an expanded grip. The blades are straight along the upper side and rounded at the sides and at the tip. The scene indicates that Classic Maya paddles

Figure 7.1. Paddler gods from incised bones, Tikal's Temple 1 burial 116. By Mary Lee Eggart.

were held by the shaft, with one hand near the top and the other hand above the blade. Other artistic depictions of canoeists paddling are known from Piedras Negras, recording a Yaxchilán emissary's trip downstream to attend a ruler's accession to the throne (Marcus 1983; Thompson 1951; Trik 1963).

The canoe paddle was discovered at the K'ak' Naab' site. The sea floor at K'ak' Naab' is firm mangrove peat overlain by 10–15 cm of loose silt. Pottery sherds extend over an area of about 10 × 15 m and are embedded in the mangrove peat, with some protruding into the overlying silt. Wooden posts were mapped at the site, demarcating a structure (McKillop 2007b). The paddle blade protruded from the mangrove peat into the overlying silt. The edge of the paddle blade in the silt was worm-eaten where it extended from the mangrove peat. The opposite side of the blade was broken.

The K'ak' Naab' paddle shows the actual size of paddles used by the Late Classic Maya: The paddle is 1.43 m in length with a round shaft that is 5 cm in diameter (Figure 6.13). The grip of the K'ak' Naab' paddle is rounded and smooth, with flaking scars visible, such as could have been produced by a chert adze recovered from the site. The upper edge of the blade flares at a 90-degree angle to the shaft. The blade extends 8 cm from the shaft on one side, but only 2 cm on the other side, where it was broken in antiquity. The blade of the paddle is rounded at the tip. A raised area on one side of the blade, 1.40 cm in height, continues the shape of the shaft. Pictorial depictions incised on bones from burial 116 in Temple 1 at Tikal (Trik 1963) show paddler gods with two hands on the shaft of the paddle in positions that would not have been effective for paddling an actual canoe (McKillop 2017b; McKillop et al. 2014). In order to gain leverage to move the canoe, the paddle needs to be held at the grip and lower at the juncture of the shaft and blade. The position with which the paddler gods are holding the paddles indicates the artist was unfamiliar with paddling a canoe, but the paddler gods were supernatural and not realistic depictions (see Trik 1963).

A modern replica made of sapodilla with both sides of the blade was paddled in Joe Taylor Creek and found to be a good paddle from steering at the stern and basic paddling from the bow of a canoe. The heavy weight of the sapodilla wood is not relevant if the paddle is kept in the water during paddling, doing the Indian stroke, for example. The long shaft of the paddle is suitable for seafaring canoes that have high sides to fend off waves but also good for smaller canoes with lower sides. The small blade is good for long paddles since a large blade would tire the paddler.

The paddle was carved from a species of *Manilkara*, probably *M. sapote* in the family Sapotaceae (R. Miller, personal communication 2004; McKillop 2005a). Although waterlogged, the wood is fresh in appearance, preserving the original light-brown color of the wood. Sapotaceae wood also was identified from Late Classic midden deposits at the nearby island community of Wild Cane Cay (McKillop 1994). A wooden spear from Actun Polbilche cave, Belize, was made from *M. sapote* (Pendergast 1974). Although *M. sapote* is better known for its sap, which provided chicle for chewing gum, the wood is a durable hardwood. Deciduous hardwoods do not grow in the mangrove ecosystem surrounding the lagoon or on the adjacent pine savanna. However, they form the deciduous rainforest south of Punta Ycacos Lagoon, along the Deep River (Wright et al. 1959).

There are artistic representations of canoes on murals at the Temple of the Warriors at Chichén Itzá (Thompson 1951:Figure 1), painted pots, incised bones from burial 116 from Temple 1 at Tikal (Trik 1963:Figures 3a, 4–7), and a gold disk from the Cenote of Sacrifice from Chichén Itzá (Coggins and Shane 1984). Canoes are depicted on a wooden lintel from Piedras Negras recording a trip from Yaxchilán to attend a dynastic accession ceremony (Marcus 1983:Figure 9). Miniature models of boats, similar in shape to the pictorial depictions, have been recovered from several Classic Maya sites, including canoes carved from manatee ribs from Altun Ha and Moho Cay (McKillop 1984:Figure 4; McKillop 2002:Figure 3.38a; Pendergast 1979:Figure 46b) as well as examples in clay from Stingray Lagoon (McKillop 2002:Figure 3.38b), Orlando's (McKillop 2002:Figure 3.38c), and Site 74 at the Paynes Creek Salt Works (Figure 6.16).

The Role of Wild Cane Cay in the Salt Trade

Wild Cane Cay survived as an autonomous trading port when inland communities of Nim Li Punit, Lubaantun, Pusilha, and others in southern Belize were abandoned at the end of the Classic or Terminal Classic periods (see Figure 1.2). The Postclassic witnessed exponential growth of sea trade, with Wild Cane Cay figuring as a major player linking external worlds and the Maya (McKillop 1989, 2005b, 2016b, 2017b). Opportunistic Maya at the trading port of Wild Cane Cay, accustomed to trading with their inland neighbors for salt and other marine resources, were able to build new trading alliances with emerging politics far to the north, tying into the circum-Yucatán sea trade of the Postclassic. The Maya on Wild Cane Cay realigned their trading partners with more distant areas in Honduras, Guatemala, northern Belize, the Yucatán, and central Mexico.

A Postclassic coral building foundation from Fighting Conch mound at Wild Cane Cay was dedicated with the sacrifice of a young woman accompanied by an imported Las Vegas Polychrome pottery vessel from Honduras (Joyce 2018; McKillop 2005b:Figures 6.8–6.9). Obsidian from the La Esperanza outcrop in Honduras was commonly traded outside the Maya area, so its presence at Wild Cane Cay further underscores external ties beyond the Maya area (McKillop et al. 1988). Trade with Guatemala included obsidian from the Ixtepeque outcrop as well as minor quantities from El Chayal and Rio Pixcaya, along with Tohil Plumbate pottery from the Pacific coast. Wild Cane Cay was linked to other coastal trading ports farther north, such as Marco Gonzalez and San Juan on Ambergris Caye and Isla Cerritos. They included similar repertoires of trade goods, notably Tohil Plumbate, green obsidian, Tulum (or "Payil") Red pottery, and Silho Fine Orange. The high quality and similar tool styles point to Colha in northern Belize as the source for chert at Wild Cane Cay (and the Paynes Creek Salt Works).

Wild Cane Cay figured in Classic period trade along the coast and contributed to the inland trade of commodities from the sea. Production and distribution of salt at the Paynes Creek Salt Works by surplus household production may have defined the relations of the coastal Maya with inland polities whose inhabitants needed salt. Because of the distance to the coast and the skills needed for salt production and canoe navigation, the Maya at inland centers may have found it cost effective to obtain salt through marketplace trade at their communities. As farmers who are busy in the rainy growing season, they could have traveled to the coast to make salt in the dry season, but why would they when there were salt producers bringing salt, salted fish, and other maritime commodities to inland markets? Surplus household production focused on different commodities on the coast and inland. Salt traders may have returned with inland corn as ballast in their canoes that would also enhance their diet, heavy on fish and tree crops (McKillop 1994, 1996b, 2005b). Uxbenka and Lubaantun were small centers, with estimated populations of about 3,500 and 3,000 people, respectively (Jordan and Prufer 2017), and Nim Li Punit also was a small community in contrast to cities such as Caracol, with 110,000 people (Chase et al. 2015).

Understanding the role of Wild Cane Cay in the Classic period salt trade is dependent on defining whether the Paynes Creek salt workers lived in nearby residences, at Wild Cane Cay, or elsewhere. Wild Cane Cay began as a fishing community at the end of the Early Classic, became a coastal trading port in the Late Classic, and expanded in the Postclassic. The Classic period archaeological deposits are deeply buried below the water table by Postclassic deposits and associated stone architecture. Therefore, the Classic deposits are only known from stratigraphic excavations below the water table in two excavations. Those excavations yielded abundant well-preserved fish and other animal bones, plant food remains, pottery, and chert and obsidian artifacts as well as burials (McKillop 1994, 1989, 2005b; McKillop et al. 1988), which are good evidence of residential household deposits indicating settlement contemporary with the use of the Paynes Creek Salt Works. It was a short paddle to the Paynes Creek Salt Works from Wild Cane Cay, making living on Wild Cane Cay a likelihood for some or all of the salt workers. The natural harbor at Wild Cane Cay may also have served as a staging area and even storage location for boatloads of salt, cacao, and dried fish.

Spatial patterning of nonbriquetage artifacts at Paynes Creek Salt Work 7 provide tantalizing clues that one or more of the 10 wooden buildings at the site may have been residences (see Figure 6.2). Three buildings lack briquetage, including a building associated with each of the three plazas at the site. The presence of nonsalt artifacts is further evidence of residential use of some of the buildings at Paynes Creek Salt Work 7. Household activities include spinning cotton, as indicated by a worked spindle whorl; fishing, as indicated by a stone net weight and by use-wear on chert tools; and ritual, as indicated by a vase with an incised glyph panel below the rim, a snuff bottle, large incense burners, small *candelero*-type incense burners, and Belize Red serving vessels.

Ek Way Nal (Site 60) may also have had residences in addition to salt production. Ek Way Nal is a large site with many buildings whose outlines are obscured by the dense distribution of artifacts on the sea floor and by the difficulty of the field team working in deep water without dive equipment. A number of artifacts point to domestic, residential activity that also suggests Ek Way Nal was a residence for salt workers that predates Site 7. The abundance of Lubaantun-style figurine whistles and Warrie Red unit-stamped designs similar to Lubaantun (Hammond 1975; McKillop 2002) contrasts sharply with their lack at Site 7. Site 7 has an abstract unit-stamp motif and other plain Warrie Red jars but lacks the S, reverse S, naturalistic monkey motifs, among others (see Figure 6.3).

The Role of Cacao in the Salt Trade

Cacao and salt are linked as commodities for production and trade by the abundance of Late Classic ocarinas at the Paynes Creek Salt Works and Village Farm, a farming village located on the mainland coast of Port Honduras between Middle River and Seven Hills Creek. The soils support a large, organic cacao farm, highlighting the likelihood that chocolate was also grown there in antiquity. Stratigraphic excavations in 2003 by the author revealed Late and Terminal Classic midden deposits with ocarinas and unit-stamped pottery similar to pottery from the Paynes Creek Salt Works. Ocarinas and unit-stamped pottery elsewhere at the site expands the assemblage. Obsidian was assayed to El Chayal using a portable X-ray fluorescence (XRF) machine. Some of the salt workers may have lived at Village Farm and commuted daily or periodically to the salt works (Table 7.1). In any case, Village Farm is en route from the salt works to inland communities, especially if travel near the mainland is selected over rough waters between the ranges of offshore cays.

Trade and Other Geopolitical External Relations

The presence of trade goods at the workshops helps to identify consumers of the Paynes Creek Salt Works salt as well as to reconstruct the trade relations and role of the salt works in the economy of southern Belize and beyond. The presence of different trade pottery at the salt works also can help identify consumers of salt and the kinds of integration in the economy of inland southern Belize. The only pottery made locally was the briquetage and other Punta Ycacos Unslipped pottery, made from quartz sand and clay locally available in the lagoon and cut banks along the shore of the pine savanna. Artifacts were studied at the Lagoon Lab and at labs at Village Farm, including use of a 3D scanner and portable XRF machine.

Trade pottery includes Belize Red dishes identical to Belize Red pottery from the Belize River valley (Chase and Chase 2012; Gifford 1976), Warrie Red unit-stamped pottery with motifs similar to pottery from inland sites in southern Belize (McKillop 2002), and ocarinas (figurine whistles) similar to those at Village Farm and inland sites such as Lubaantun as well as a variety of other nonlocal pottery. All calcite-tempered pottery was made elsewhere and transported to the salt works by the workers or traded. Calcite-tempered pottery includes Mangrove Unslipped jars used for storing brine and loose salt as well as a variety of unknown pottery types. Mapping the spatial distribution of trade and nonlocal Belize Red, Warrie Red unit-stamped pottery, ocarinas, and obsidian can help identify whether certain workshops were the main recipients of trade items. Mapping the distribution of distinctive motifs on unit-stamped pottery may identify temporal or spatial patterns of distribution. Identification of the possible origins of the pottery by thin-section petrography helps reconstruct distribution of salt to inland markets.

Belize Red Pottery

Belize Red pottery includes sherds from shallow dishes with a basal break, including some with a notched angle. Belize Red is a red-slipped, volcanic ash-tempered ware found at sites in the Port Honduras–Paynes Creek region of the coast of southern Belize (McKillop 2002), originally named in the Belize Valley (see Gifford 1976). Belize Red also is present inland in southern Belize at Lubaantun (Hammond 1975). Distinctive incised decoration on the exterior includes horizontal lines below the rim and

Table 7.2. Belize Red pottery traits by area

Area	Notched basal angle	Plain basal angle	Diagonal incised lines	Horizontal incised lines	Ridge	Scallop	Total
East Lagoon	6	—	1	5	—	—	12
Main Channel	3	8	7	8	—	—	26
West Lagoon	1	1	—	3	1	—	6
West Point	1	2	—	—	—	—	3
Stingray area	—	1	1	1	—	1	4
Total	11	12	9	17	1	1	51

above the basal break of open bowls or dishes. Some vessels have diagonal incised lines on a panel above the basal break on the exterior.

Belize Red pottery is overrepresented in the survey collection. The sherds—even sherds without identifiable form—were mapped and collected on survey since they are a Late to Terminal Classic temporal marker, so the presence of this pottery type in the survey collection is clearly overrepresented compared to other pottery types. Belize Red pottery has a visibly distinctive red color that often remained on the pottery, in contrast to the lack of preservation of paint for other pottery (such as Warrie Red). Even the feel of the pottery, with a soft and smooth surface that is easily scratched by a fingernail, makes Belize Red an easy pottery type to identify underwater without even seeing the sherd.

Belize Red pottery is widely distributed at underwater sites in the East and West Lagoons, Main Channel, and West Point. Rims were found in about equal numbers in the East and West Lagoons. The sample of 63 Belize Red sherds includes 16 from the East Lagoon, 18 from the Main Channel, 2 from Stingray Lagoon area, 15 from the West Lagoon, and 12 from the West Point area. The sherds represent 43 serving dishes, 3 cylinder vessels with nubbin or slab feet, 8 small pedestal-based vessels, 2 ring bases, and 9 vessels of unknown shape. Variation in the Belize Red dishes includes the presence of notched or plain basal angles (Table 7.2). There were 11 notched basal angle bowls and 14 plain basal angle dishes as well as 1 basal ridge. Six of the notched basal angle rims are in the East Lagoon, with only one in the West Lagoon. The widespread distribution of Belize Red underscores marketplace access and further supports the model of surplus household producers working at the salt kitchens and transporting their salt to market.

Dishes often have an incised panel decoration on the exterior between the rim and basal angle. The panel is defined by a circumferential incised line below the rim and one to two additional incised lines above the basal angle, sometimes with a diagonal, incised, or grooved line or lines in the panel. The nine rims with diagonal incised lines all have plain basal angles. There are 19 with horizontal incisions, including nine with plain basal angle and two with notched basal angle. One rim sherd has a gouge-incised, scallop pattern repeated in the panel. The horizontal, diagonal, and gouge incisions are mainly in the East Lagoon and Main Channel, with few in the West Lagoon and Stingray Lagoon. If the occurrence of plain versus notched basal angle types of panel incision on the exterior of bowls are temporal markers, the distribution of all styles across the lagoon system indicates settlement developed in all areas instead of beginning in one area and spreading across the lagoon system.

In a sample of Belize Red dishes from burials at Caracol, Arlen Chase and Diane Chase (2012:Figure 2) identify vessels with round bases as Late Classic and flat bases at Terminal Classic (Chase and Chase 2012:Figure 3). They describe variability in vessel forms in the early Late Classic, with footed dishes and plates and with large bowls and cylinder vessels. Belize Red was more standardized by the end of the Late Classic, featuring sag-bottomed dishes with hollow, bulbous feet (Chase and Chase 2012:Figure 2g). The Terminal Classic included large footed dishes with flat bases (Chase and Chase 2012:Figure 3d) and smaller footed bowls, including some with notched flanges (Chase and Chase 2012:Figure 3b).

The West Point sites have lots of Belize Red pottery, but it generally lacks decoration, making it different from other areas at the salt works. The West Point sites lack the diagonal and horizontal incisions that form a panel on the exterior of Belize Red open serving dishes. They are more common in the Main Channel and East Lagoon and present in the West Lagoon. The West Point area has only one notched and two plan basal angles on Moho Red dishes. Notched angles are common in the Main Channel and East Lagoon. West Lagoon sites have few of these features.

Unit-Stamped Pottery

Tinaja Red is a common red-slipped ceramic group dated to the Late and Terminal Classic in the Southern Maya Lowlands. However, the decorative technique of unit-stamping on Tinaja vessels has a restricted distribution, being most common in southern Belize, adjacent Guatemala, western Belize caves, and Caracol but in limited quantities elsewhere, including Tikal and Barton Ramie (Kosakowsky et al. 2013). The unit-stamped pottery at the Paynes Creek Salt Works and elsewhere on the coast and cays of southern Belize in Port Honduras has eroded surfaces, so the slip is rarely visible. The acidic mangrove peat has dissolved the calcium carbonate in much of the pottery, including unit-stamped pottery (McKillop et al. 2018). It is referred to as Warrie Red at the Paynes Creek Salt Works and elsewhere on the nearby coast and cays due to its eroded surfaces but also because of the variety of stamped motifs that have more diversity than other Belizean sites and more naturalistic motifs than those in the Petexbatun (McKillop 2002:77–86, Figures 3.27–3.30).

First identified at Altar de Sacrificios as Pantano Impressed Stamped Variety in the Tinaja group (Adams 1971), unit-stamped pottery is also reported from Ceibal as Chaquiste Impressed Stamped Variety in the Subin group (Sabloff 1975). At Lubaantun, unit-stamped pottery is included in the Remate Red group (Hammond 1975). Unit-stamped pottery also is found in caves and other sites in the Maya Mountains, including Las Cuevas (Kosakowsky et al. 2013) and various caves investigated by David Pendergast, including bowls (Pendergast 1971:Figures 9, 10, 11a–j) and jars (Pendergast 1971:Figures 13d–q, 14a, 14g–k) at Eduardo Quiroz Cave and other pieces at Actun Balam cave (Pendergast 1969:Figures 8c, 8e, 9d, 9h) and Rio Frio Cave (Pendergast 1970:Figure 11o). At Las Cuevas, unit-stamping comprises 10 percent of the ceramics, mainly on incurring monochrome red bowls but also on jars. At Las Cuevas, unit-stamped vessels focus on comb-stamping and S unit-stamp motif but lack abstract motifs of the Petexbatun or the animal motifs of Lubaantun and the southern coast (Kosakowsky et al. 2013:29–30, Figures 3–4). Hammond (1975:305) suggests there were two main areas of unit-stamping, including a western area including the Pasion area of Ceibal and Altar de Sacrificios and nearby sites as well as an eastern area including the upper Belize Valley, Mountain Pine Ridge, and the Lubaantun area. The Paynes Creek stamps include a diverse array of designs, with most similarities to Lubaantun, including comb, S and reverse S stamps, monkey stamps, and a variety of abstract motifs (McKillop 2002:85).

Warrie Red pottery is distinctive due to its smooth surfaces, traces of red slip on the exterior of some sherds, and the unit-stamped decoration around the shoulder of jars. Unit-stamp decorations are made by impressing the same design repeatedly around the shoulder of the vessel. A unit-stamp decoration occurs alone, with another unit-stamp, or with impressed or incised decorations. The unit-stamp decorations include a forward S, reverse S, naturalistic or abstract monkey, comb, peanut, and abstract designs (Figure 7.2). The designs co-occur with one another in various combinations. Impressed decoration includes round and teardrop punctate and thumbnail and one or more circles. A combination of three circles occurs on several sherds. Incised decoration includes a circumferential line on the shoulder at the juncture with the vessel neck as well as horizontal or vertical lines elsewhere on the shoulder, sometimes associated with circles. The widespread access suggests marketplace trade for Warrie jars. The variation in decorative unit-stamp motifs as well as the different pastes identified by Linda Howie suggests the jars were obtained from different marketplaces to which the salt workers brought their salt, cacao, salted fish, and marine resources to trade (McKillop et al.). Petrographic analysis of thin sections of 15 Warrie Red sherds revealed 13 different fabrics that do not correspond with specific unit-stamp motifs. Most of the ceramic fabrics can be linked geologically to localities in southern Belize. Mineralogical and other compositional differences suggest different production locales.

All decorated sherds as well as large rims were collected, resulting in overrepresentation of Warrie Red pottery in the collections. In addition to the individually mapped Warrie sherds recorded by letters, Warrie sherds collected by site in 2004 and 2005 were added to the mapped data. The 2004 and 2005 sherds were recorded by artifact numbers given to the pottery when it was drawn.

Of the sample of 210 motifs on Warrie Red pottery in the five areas of the Paynes Creek Salt Works, there are some patterns evident by cross-tabulating location with motif (Table 7.3). The common decorative motifs at West Point sites are incised lines and punctate, also common at West Lagoon sites. West Point sites lack comb-stamping, which is common at the East Lagoon and Main Channel sites. West Point sites lack the S motif and only have one monkey stamp sherd, which is more abstract than the monkey motifs elsewhere. Both S and monkey stamps are common in the East Lagoon and Main Channel. Most unit-stamp motifs are on Warrie pottery in the East Lagoon and Main Channel, including all comb motifs, most S and reverse S motifs, and more monkey and abstract motifs. The West Lagoon and West Point sites have more circle and punctate motifs.

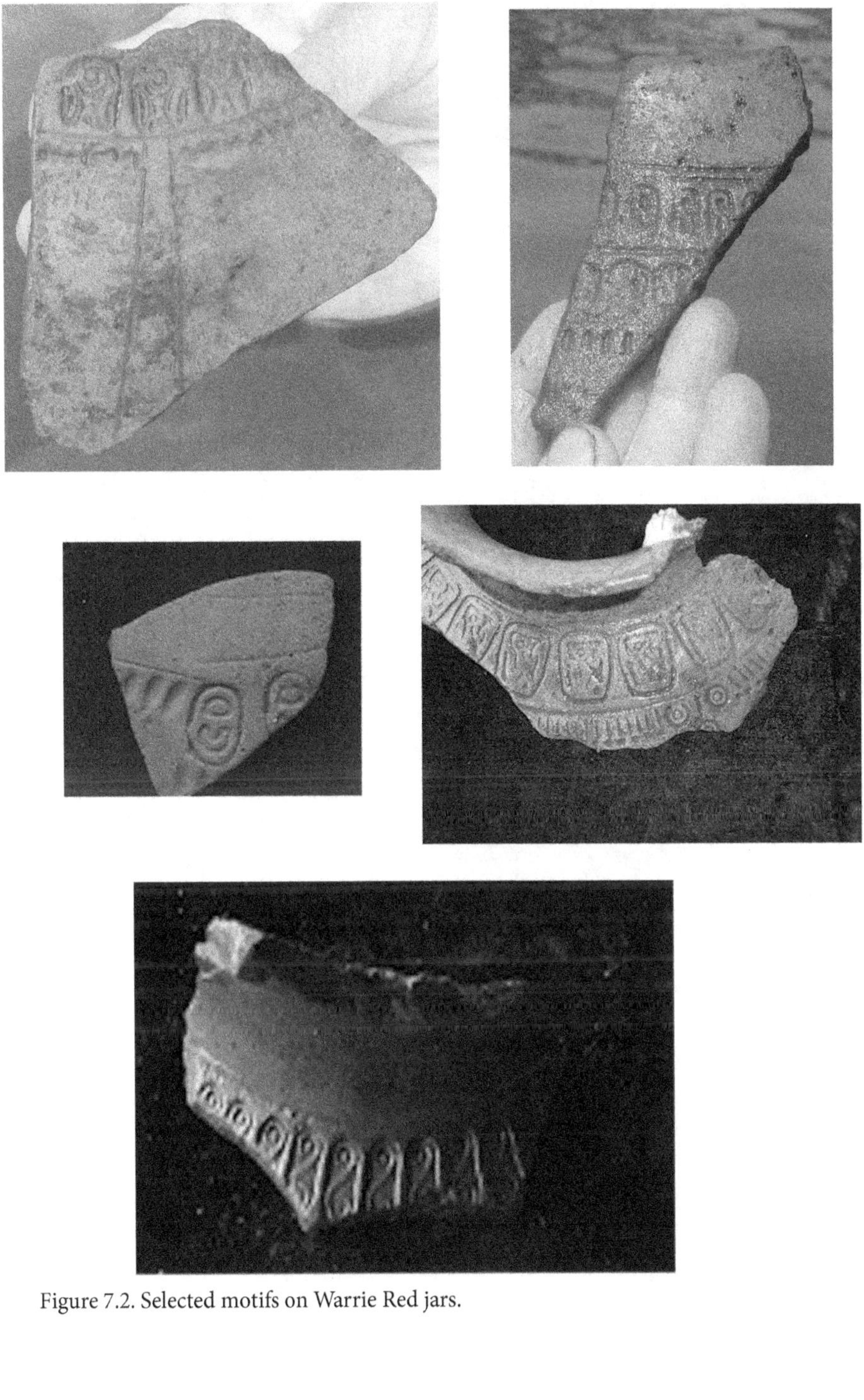

Figure 7.2. Selected motifs on Warrie Red jars.

Table 7.3. Warrie Red motifs by area

Area	S	Reverse S	Monkey	Comb	Circle	Line	Abstract	Thumbnail	Punctate	Total
East Lagoon	5	6	2	8	2	7	5	—	7	42
Main Channel	6	4	3	22	2	11	5	1	7	61
West Lagoon	2	—	2	—	7	21	3	2	18	55
West Point	—	2	1	—	3	19	2	1	14	42
Stingray area	1	—	—[a]	—	—	1	—	—	1	3
Total	14	12	8	30	14	59	15	4	47	203

[a] Monkey motif was previously found at Stingray Lagoon site (McKillop 2002:Figure 3.27).

Although Warrie Red motifs are distributed in all areas of the Paynes Creek Salt Works, there are concentrations of specific motifs in particular areas or salt works. The East Lagoon has few mapped Warrie sherds but most of the numbered Warrie sherds, a result of the focus on underwater survey in the East Lagoon in 2004 and 2005. The S and reverse S motifs are more common in the East Lagoon and Main Channel than elsewhere. The monkey motif is rare, being found only at four sites in the East Lagoon / Main Channel (Sites 12, 30, 60, 61) and two sites in the West Lagoon / West Point (Sites 74 and 7). The comb motif is found in the East Lagoon / Main Channel. The abstract "bottle cap" unit-stamp motifs are rare, being found in the East Lagoon / Main Channel (Sites 29, 60, and 61) and West Lagoon / West Point (Sites 3, 48, 103). The impressed circle and incised lines are found more frequently at sites in the West Lagoon, West Point, Stingray, and Main Channel than in the East Lagoon, where they are rare. The Warrie Red jars with circle and line motifs also tend to be thicker-walled and heavier than the Warrie jars with other motifs.

Other Pottery

Some pottery found on sea floor survey points to a long duration of trade from the Early Classic to the Terminal Classic. Early Classic pottery includes Dos Arroyos Orange Polychrome bowl sherds from the Eleanor Betty site. Late Classic pottery includes a vase rim with a painted glyph on the exterior below the rim from the Eleanor Betty site, a Paradero Fluted partial vessel from the John Spang site (Site 72), and a small bottle from Site 7 (see Figure 6.3b). Terminal Classic pottery includes a partial vase with a panel of incised glyphs below the rim on the exterior from Site 7 (see Figure 6.3a), a globular tan-colored vessel fragment with wide flutes from Site 69, and a globular vessel with a gouge-incised panel on one side from the Carpenter site (see Figure 6.3e). In addition, there are several "fat-lipped" Terminal Classic basin sherds similar to pottery from the Sibun River valley and northern Belize (Harrison-Buck and McAnany 2013; Masson and Mock 2004). Other nonlocal pottery includes a prong from an incense burner similar to examples from the Belize Valley.

Obsidian

A comparison of the salt makers' obsidian source use with other nearby coastal and inland communities indicates the Paynes Creek salt makers were tied to the coastal trade route dominated by Wild Cane Cay. For the 40 pieces of obsidian from survey that were chemically characterized using a portable XRF machine (Sills and McKillop 2018b), 23 were identified to the Ixtepeque source, 15 to El Chayal, and 2 could not be identified to a source (Table 7.4). Differences in source use between sites in the East and West Lagoons reflect differences identified by modal analysis for Belize Red and Warrie Red pottery (Table 7.4). The West Point sites are distinct from the other areas in several ways and may be exclusively Terminal Classic in age. The lack of ocarinas at the West Point sites may indicate they are later than the East Lagoon and Main Channel sites, where ocarinas are common. Obsidian source use is mainly Ixtepeque at the West Point sites, typical elsewhere of the Terminal Classic. A Terminal Classic glyph vessel as well as wedges near the surface of the sea floor (not buried by mangrove accretion) suggest Site 7 in the West Point area is a more recent site than the East Lagoon sites.

Obsidian source use at the Paynes Creek Salt Works resembles the pattern at nearby Wild Cane Cay for the Late and Terminal Classic periods (McKillop et al. 1988). The Late and Terminal Classic period at Wild Cane Cay has 15 Ixtepeque obsidian, 12 El Chayal obsidian, and 2 Source Z obsidian items. The Early Postclassic at Wild Cane Cay witnessed an increase in the use of Ixtepeque obsidian as well as an increase in the diversity of obsidian sources used, with seven sources identified. The Early Postclassic included 65 Ixtepeque, 6 El Chayal, and 1 blade each of Source Z and Rio Pixcaya. An additional sample assigned to the Early Postclassic included 1 Ucareo, 1 La Esperanza, and 3 Source Z. Source Z may be located along the Rio Motagua, upriver from the Puente Chetunal (McKillop et al. 1988). Source Z also was used by the Classic period Maya at Quirigua, a city along the Rio Motagua.

Comparison of obsidian from the Paynes Creek Salt Works with other nearby sites dated to the Classic period generally shows sites in the interior relied more on El Chayal, whereas coastal sites relied on Ixtepeque (Table 7.5). Both Wild Cane Cay and the Paynes Creek Salt Works had more Ixtepeque than El Chayal obsidian, but still had a significant amount of El Chayal. The Classic period obsidian samples (AD 300–700)

Table 7.4. Obsidian source use by area at the salt works

Area	El Chayal	Ixtepeque	Unknown	Total
East Lagoon	1	—	—	1
Main Channel	2	2	—	4
West Lagoon	6	8	2	16
West Point	2	12	—	14
Stingray Area	4	1	—	5
Total	15	23	2	40

Table 7.5. Obsidian source use at selected southern Belize sites

Source	Paynes Creek Salt Works	Wild Cane Cay	Uxbenka (AD 300–900)	Ek Lux (AD 300–700)	Pusilha (Early Classic)	Pusilha (Late Classic)
El Chayal	38%	42%	85%	83%	94%	92%
Ixtepeque	58%	52%	11%	17%	6%	8%
Source Z		7%				

from Uxbenka and Ek Lux in the interior of southern Belize have mainly El Chayal obsidian, with 85 percent and 83 percent, respectively. Farther south, Late Classic (AD 600–900) obsidian from Pusilha has even more El Chayal, at 95 percent, similar to the Early Classic, at 94 percent (Daniels and Braswell 2014). Overall, there is more use of Ixtepeque along the coast versus greater use of El Chayal inland during the Classic period. The similar patterns of obsidian source use at Wild Cane Cay and the Paynes Creek Salt Works suggest that the salt workers obtained obsidian from the cay instead of from inland trade and that the salt makers may in fact have lived on the cay.

Jadeite and Other Greenstones

Jadeite and other greenstones made into axes, adzes, and gouges were traded from the jadeite-producing area along the Motagua River for use at the salt works (McKillop et al. 2019). The jadeite gouge with the rosewood handle from the corner of a salt work building at Site 60 is high-quality jadeite with few impurities. Stone axes from various salt works also were made of jadeite.

Chert

The types and distribution of chert stone tools and flakes indicate that tools were not produced at the salt works. The chert resembles northern Belize chert tools as well as with those from Wild Cane Cay. Chert is not locally available in the coastal lagoon area although there are outcrops of dolomite and chert nearby at the karst hills north of Punta Gorda. Generally, the chert in southern Belize consists of small, friable pieces of poor-quality material as well as fist-sized river cobbles that are not suited for stone tool production of the quality of objects found at the salt works. Chert tools likely were traded to the nearby trading port of Wild Cane Cay from high-quality, northern Belize chert outcrops at Colha and surrounding area (McKillop 2005b). The 62 chert stone objects individually mapped include chopper pounders (n = 8), unifacial stemmed blades (n = 15), shouldered thin bifaces (n = 8), large blades (n = 3), adzes (n = 12), biface blades (n = 1), bifaces (n = 5), flakes (n = 6), and other objects. Imported chert tools were not common but were widely distributed at the salt works, underscoring the marketplace access (McKillop and Aoyama 2018).

Ritual and Political Economy of Salt Addiction

Discussions of the settlement history and geopolitics of southern Belize focus on inland cities, largely ignoring the long trajectory of Preclassic through Postclassic settlement on the coast and offshore cays. By implication, the coast was not important in the geopolitics of ancient Maya society in southern Belize. Alternatively, external relations of the inland Maya were not directed toward the coast or sea trade but instead toward other inland politics so that any trade with the coast was marginal to the geopolitical world of the inland Maya. Coastal sites may have been marginalized because they lack dynastic records. Furthermore, sea-level rise has inundated the coast and its ancient settlements, obscuring their visibility in the modern coastal landscape. Still, the coast of southern Belize was a source of ritual and dietary resources for the inland Maya as well as a transportation avenue for goods and resources from farther away. What were the relations between the coastal and inland Maya in southern Belize? Did the inland Maya of southern Belize figure in the political economy, settlement history, and geopolitics of the ancient Maya on the coast and cays of southern Belize? Were the inland Maya of southern Belize marginal to the geopolitical world of the coastal Maya of southern Belize?

Early Middle Preclassic settlement in southern Belize is documented by Jocote Orange-Brown pottery in a shell midden at Ich'ak'tun (formerly called Butterfly Wing site), located on the south side of the Deep River mouth (see Figure 1.2; McKillop and Robertson 2018). External connections at this early time period are noted by an obsidian flake from the Ixtepeque source. Jaime Awe and Paul Healy (1994) document the widespread production of obsidian flakes from cobbles beginning in the Middle Preclassic period and prior to the core-blade technology typical from Late Preclassic times onward.

The Preclassic period witnessed settlement throughout the Maya Lowlands, with the Sierra Red pottery marking broad patterns of trade and communication during the Late Preclassic, including southern Belize, as noted by inland settlement at Uxbenka (Prufer et al. 2007) and coastal settlement at Ich'ak'tun (McKillop and Robertson 2018) and Nim Li Punit during the Protoclassic, about AD 150 (Prager and Braswell 2016). Uxbenka began as a small farming community located on good arable land. Ich'ak'tun site followed pan-lowland traditions of Protoclassic mammiform tetrapod pottery vessels, Sierra Red pottery, and an obsidian flake industry. Ich'ak'tun also followed the Preclassic tradition of shell middens, which continued later in the area at the Schmidt site and other submerged salt works in deep water in the Main Channel at the Paynes Creek Salt Works (Feathers and McKillop 2018; McKillop 2017a). Settlement expanded in the Early Classic, both at inland farming communities and on the coast at Wild Cane Cay and Pelican Cay.

Settlement in southern Belize increased on the coast and inland throughout the Classic period. The rise of dynastic polities at Uxbenka, Nim Li Punit, and Pusilha accompanied population increase in southern Belize both inland and on the coast. Dated monuments record dynastic histories at Uxbenka, Nim Li Punit, and Pusilha (Braswell et al. 2004, 2007; Wanyerka 2005). Ceramic chronologies place Lubaantun in the Late Classic Tepeu 2 and 3 (Hammond 1975) and extend settlement near Pusilha to the end of the Late Classic during Tepeu 3 (Bill and Braswell 2005). Nim Li Punit had limited external connections until about AD 800, with the appearance of Belize Red pottery (Prager and Braswell 2016:276). Tomb 5 in Structure 3 had Terminal Classic polychrome and Belize Red vessels, a chert eccentric, and a carved jadeite plaque with hieroglyphic inscriptions (Prager and Braswell 2016).

A major factor in coastal–inland relations during the Classic period was the inland demand for salt to meet the basic daily biological needs of the growing population. In addition to salt, salted fish and other marine resources—for both ritual and subsistence uses—forged inland interest in the coast. The importance of the coast as a transportation gateway for goods and resources from outside the region may also have fueled interest by the inland dynastic Maya on the adjacent coast. The presence of Colha chert on Wild Cane Cay and Village Farm and in minor amounts at the Paynes Creek Salt Works argues for canoe transport along the coast from northern Belize high-quality chert outcrops.

The earliest radiocarbon dates from Wild Cane Cay place initial settlement during the Early Classic, along with nearby Pelican Cay (McKillop 2002:158). Wild Cane Cay was first settled in the Early Classic as a fishing community. Ceramic chronology and radiocarbon-dated midden deposits at coastal communities indicate expansion of settlement in the Late Classic at Wild Cane Cay, Pork and Doughboy Point, Village Farm, Green Vine Snake, Frenchman's Cay, and the Paynes Creek Salt Works (McKillop 1996a, 2002, 2005a, 2005b, 2017b; McKillop and Sills 2016, 2017; McKillop and Winemiller 2004; McKillop et al. 2004; Somers 2007).

During the Late Classic, Wild Cane Cay became a trading port. Its rise to prominence, its proximity to the Paynes Creek Salt Works, and the inland need for salt may have allowed the island to control the salt trade (McKillop 2005a, 2005b). As a trading port, Wild Cane Cay funneled inland various maritime resources, such as stingray spines, conch shells, and seafood. The abundance of obsidian at the island underscores the islanders' easy access to this imported material and its participation in long-distance sea trade during the Late Classic.

Major factors during the Postclassic period were the abandonment of most inland cities in the Southern Lowlands; the rise to prominence of Chichén Itzá and, later, Mayapán; and Chichén Itzá's and Mayapán's participation in a much larger geopolitical world from central Mexico to Honduras (Chase and Rice 1985; Sabloff and Andrews 1986; Smith and Berdan 2003). Coastal trade figured prominently in this political landscape. Coastal settlement increased in the Postclassic, perhaps in response to population movement from inland areas, but clearly as a response to an increase in sea trade. Prominent Classic period cities in southern Belize were virtually abandoned, including those cities with public dynastic records carved on stelae at Nim Li Punit, Uxbenka, and Pusilha as well as those without—notably, Lubaantun. Apart from limited Postclassic settlement noted near Pusilha (Braswell et al. 2007), available data indicate the inland area was largely depopulated. On the coast of southern Belize, the Paynes Creek salt industry collapsed for lack of an inland market, but its port on Wild Cane Cay survived and became a major player in Postclassic sea trade and the Postclassic Mesoamerican world (Joyce 2018; McKillop 2005b). Settlement continued elsewhere on the south coast of Belize at Frenchman's Cay and Foster Farm, but there was a diminution of settlement on the coast compared to the Classic period.

Sixteenth-century Spaniards and later British buccaneers and log workers encountered Maya both on the coast and in the interior of southern Belize (McKillop 2005b:194–196; Wilk 1987). The sixteenth-century Spaniards disrupted a thriving circum-Yucatán canoe trade that included Wild Cane Cay. The presence of Spanish artifacts at Wild Cane Cay, such as fragments of olive jars stylistically dated between 1550 and 1770, indicates early historic European presence in the coastal waters north of Punta Gorda. Among the voyages of pirates, the famous British buccaneer Bartholomew Sharp captured a Spanish Dominican friar and several Maya south of Punta Gorda in 1677 (McKillop 2005b:195). According to Richard Wilk (1987), the Maya were dispersed in the rainforest in small communities inland in southern Belize. The indigenous Maya population grew in southern Belize, with Maya fleeing forced labor and other hardships in Guatemala in the late nineteenth century. Logging along the rivers of southern Belize by 1802, along with nineteenth-century Garifuna settlement of Punta Gorda, the U.S. Confederate settlement north of Punta Gorda, and settlement of the coast and cays by fisherfolk, are indicated by historic pottery, glassware, and metal at various locations, including Wild Cane Cay, Village Farm, the Sapodilla Cayes, and Deep River (McKillop 2005b:194–196).

The geopolitics of modern southern Belize are poor indicators of the dynamics of the ancient past in the region. The once arable land on the coast and cays of southern Belize has been flooded by seawater, with ancient salt works and sites submerged below mangroves and underwater.

Politics, Economics, and Ritual

The wooden buildings and associated artifacts from ancient Maya salt works provide an opportunity to examine the political, economic, and ritual factors that linked producers and consumers in Late Classic Maya society. The salt works likely developed from surplus household production located near the natural resource they were exploiting—a supersaline salt water lagoon. Their wooden structures and the associated high densities of briquetage in contrast to low densities of residential objects indicate specialized salt production. The ceramic assemblage contrasts with the diversity of pottery vessel forms and types found at Wild Cane Cay and other nearby settlements (McKillop 2002). The salt workers presumably lived year-round at the contemporary coastal settlements nearby or possibly at one or more of the large sites at the salt works that are now submerged. They were local producers who took salt upriver and along the coast to markets.

John Watanabe (2007) suggests that early Maya leaders capitalized on existing efforts of individual householders at urban centers to produce beyond household needs in order to attract marriage partners. The expansion of surplus production would have opened a door for aggrandizing leaders because the latter were able to broker intergroup relationships in their capacity as ritual experts. The state thus became a beneficiary of this economic expansion under the guise of ritual. Such mechanisms enabled the incorporation of common folk into the political hierarchy, bringing them economic as well as ritual benefits. As to how they were played out in Maya social life, Antonia Foias (2007) discusses how feasts and other rituals served to create, maintain, and reinforce power relations during the Late Classic period.

The Maya at the Paynes Creek Salt Works likely were tied into a complex network of ritual and economic relationships with the inland Maya. The coastal Maya, with their center perhaps at the trading port of Wild Cane Cay, received goods such as ocarinas, serving vessels, and other trade pottery. The coastal Maya became invested in the ritual ideology underlying the political structure of Maya dynasties, which was enacted in feasts overseen by shaman kings. Driven by such an understanding of the world, the gods, and people's place in the world, the coastal Maya reenacted these feasts in their homes and workplaces. Rather than using military force or imposing local rulers in order to incorporate the coastal

Maya salt works into a regional market economy, the ruling elites at the inland dynastic centers maintained a regular trade in salt by means of ritually sanctified trade. Albeit on a different scale, the dynastic rulers at Caracol facilitated the distribution of "a full array of items" including nonlocal and well-crafted commodities through marketplace trade at termini groups that "served as a successful management strategy at Caracol for over 200 years" (Chase et al. 2015:248).

Materiality of Ritual Behavior

Some of the salt works hosted periodic salt rituals (McKillop 2002:95–98) such as those performed at the opening of the salt production season at Sacapulas, Guatemala (Reina and Monaghan 1981). The materiality of ritual might include a clustering of objects related to sharing food and music at key sites where the salt workers would congregate periodically—such as the beginning of each salt season, as at Sacapulas. Potential ritual objects include Belize Red serving vessels, ocarinas, *candeleros*, and incense burners. They occur in minor quantities overall at the salt works and were not used to produce salt. Belize Red serving bowls and most ocarinas were not made at the salt works but were presumably acquired through trade at marketplaces when the salt workers traveled to sell their salt. Ocarinas are concentrated at Sites 60 and 61, adjacent salt works that are at the entrance to the East Lagoon and Main Channel salt works. The Stingray Lagoon site and the David Westby site in the Stingray Lagoon area have the only other concentration of ocarinas and may have served the Stingray Lagoon area as well as the West Lagoon and West Point.

The distribution of other potential ritual objects is not concentrated. Belize Red pottery is widely distributed, albeit in small quantities, at many salt works, which can be explained by marketplace trade providing everyone equal access (Chase et al. 2015; and Hirth 2009a, 2009b). Incense burners, including the small tripod and flat-bottomed containers as well as large basin-shaped containers with pedestal bases and lids, are both found at Site 60 but are present at many other sites as well. In summary, the material evidence of clustering of ritual activities is limited to the ocarinas that are overwhelmingly at Sites 60 and 61 in the Main Channel and Site 56 in the Stingray Lagoon area.

Summary

Surplus household production provided dietary salt and salt for preserving fish, meat, and animal hides. Salt workers may have lived in residences at the large sites 7 or 60 at the salt works or elsewhere in the coastal area accessible by canoe to the salt works. Wild Cane Cay and Village Farm are contemporary sites with similar unit-stamped pottery, ocarinas, and evidence of habitation. The salt workers or their surrogates transported salt cakes, salted fish, and perhaps salted meat and hides as well as other marine resources by canoe along the coast and up rivers to nearby inland Maya marketplaces (McKillop and Aoyama 2018). The canoe paddles and canoe underscore the use of water transportation routes. The similarity of obsidian source use between the salt works and Wild Cane Cay, in contrast to nearby inland cities, adds credence to a view of a coastal focus for the salt works. Salt cakes provided a storable commodity that had further trade value as a currency equivalency. Storing salt as well as salted fish also was a good risk management practice in case of periodic food shortages.

8

Salt

A Mover and Shaker in Ancient Maya Society

As a basic biological necessity, as a flavor enhancer, as a food preservative, and as a currency equivalency, salt—in the form of salt cakes—was a common commodity in the ancient Maya economy. Salt was one of many commodities that was produced and distributed through interlocking networks of exchange in ancient Maya society.

The Salt-Making Process in Maya Society

Salt was produced in the Maya area by solar evaporation on the north coast of the Yucatán, in the Maya Highlands, and on the Pacific coast, and by boiling brine on the coast of Belize, on the Pacific coast of Guatemala, and in the Maya Highlands. Depending on the salinity of the brine, it was enriched by leaching or solar evaporation before the brine was boiled in pots over fires. Evaporating brine in pots over fires is a common technique worldwide, from ancient to modern times. The salt works are characterized by briquetage—broken pots and sometimes vessel supports for the brine-boiling process. There are variations on the technique of enriching the salt content of the brine obtained either from the sea or from salt springs.

The Paynes Creek Salt Works employed a widely used technique. Salt making at Sacapulas, Guatemala, provides a model for the salt-production process at the Paynes Creek Salt Works, with additional information from salt works in northeast Thailand (Halliwell et al. 2016) and the Philippines (Yankowski 2010). The process consists of pouring salty water through salty soil to enrich the salt content of the brine, then straining the brine into pots and transferring it to salt pots. To produce salt, the brine was boiled in pots over a fire inside salt kitchens using wood fuel. Briquetage at the Paynes Creek Salt Works includes several vessel forms, including bowls, jars, and basins, as well as clay cylinder vessel supports, sockets that were placed at the top of each clay cylinder with a concave surface where the salt pot rested, and a base where the cylinder was fitted to rest on the boiling apparatus. Clay spacers may have separated jars and basins over a fire. Spacers also are reported from the Placencia Lagoon Salt Works (MacKinnon and Kepecs 1989). Spacers have two concave surfaces and have not been found attached to other briquetage.

Earthen mounds at Witz Naab' and Killer Bee are piles of discarded soil from the brine-enrichment process (Watson et al. 2013; Watson and McKillop 2019). The process likely took place near the salt kitchens, but the earthen mounds were deflated by wave action as when the salt kitchens were submerged by rising seas (Figure 8.1). Earthen mounds are common at the Placencia Lagoon Salt Works (Sills 2016, 2017), providing a gauge for their prior ubiquity at the Paynes Creek Salt Works. The enriched brine was collected in a container below, as shown by the canoe at the Eleanor Betty site with the clay funnel below.

Similar pottery vessels were used at the salt works on the coast of China in Shandong Province (Flad 2013), at the Zhongba salt works (Flad 2011), in Niger (Gouletquer 1975), and in the Philippines (Yankowski 2010). Briquetage vessels similar to those used at the Paynes Creek Salt Works are found at other salt works, notably the Placencia Lagoon Salt Works (MacKinnon and Kepecs 1989; Sills 2016, 2017) and Zhongba in China (Flad 2011). Small open vessels were used at San Mateo Ixtatán, in the Huehuetenango department of highland Guatemala (Andrews 1983:81–87, Figures 4.6–4.10). Larger pots are reported from Salinas de los Nueve Cerros (Andrews 1983:Figure 4.13; Woodfill et al. 2015), from Ixtápa near Zinacantán in the highlands of Chiapas (Andrews 1983:59) where pots 70 cm in diameter and 90 cm high were used. The entire Pacific coast of Guatemala has briquetage salt production, based on boiling brine in large pots after the brine was enriched by pouring salty water over salty soil in old canoes (Andrews 1983:68); the Guzman Mound salt works excavations are an example (Nance 1992). Clay pots were estimated at 60–80 cm in rim diameter and 60–100 cm in height (Andrews 1983:68), identical to pots at Ocós reported by Michael Coe and Kent Flannery (1967:Plate 28K). Large metal containers replaced clay pots, which were later replaced by large industrial solar evaporation works introduced in the twentieth century

Figure 8.1. Panorama of the East Lagoon with underwater sites hidden from view.

(Andrews 1983:62–72). Major solar evaporation works are at San José Iztapa, which is the largest salt producer in Guatemala, and Champerico, which is the second-largest Guatemalan salt producer. On the coast of Chiapas, Mexico, salty water was poured over soil in raised wooden boxes, and the enriched brine was evaporated in small ponds (Andrews 1983:62–68), similar to the La Placita Salt Works in Michoacan (Williams 2003).

Clay cylinders were used to support brine-boiling pots over a fire in the Maya area and elsewhere. Clay cylinders are found at salt works on the Pacific coast of Guatemala (Coe and Flannery 1967:91, plate 28), on the coast of Belize at the Placencia Lagoon Salt Works, Northern River Lagoon and 10 nearby sites, Saktunja and 12 more sites on Midwinter's Lagoon, Moho Cay, and the Paynes Creek Salt Works. Clay cylinders are common in England and elsewhere in Europe, in Niger, and in Vietnam. Stones were used to elevate salt pots over a fire at Sacapulas. Elsewhere, clay ovens or below-ground fire pits were used in which the brine-boiling pots were nested in holes or set on top of the oven surface.

At the Paynes Creek Salt Works, salt was produced through surplus household production that can be described as contingent multicrafting since salt making was contingent on a host of other activities. Outsourcing may have included household specialization in fishing, woodworking, or salting fish, for example. Some of the contingent activities may have been outsourced to relatives or other specialists, as described with modern Maya pottery making in the Yucatán (Arnold 2017).

Marketplace Trade

Commodities were exchanged at marketplace trade and through tribute, ritualized gifting and feasting among dynastic leaders as depicted in scenes on painted pots, and more informal exchanges among households. Traci Ardren and colleagues (2016) describe cotton cloth production on the overland route from the coast to Chichén Itzá, suggesting that cotton was produced as tribute required for the Maya leadership at Chichén Itzá. Marketplace trade was a fundamental component of the Maya economy from at least the Classic period (Freidel et al. 2016; King 2015; Masson and Freidel 2012) and included the largest cities (Chase et al. 2015) and modest villages (Sheets et al. 2015). Marketplace trade took place at formal marketplaces that had stone market stalls such as at Tikal (Jones 2015) and Caracol (Chase et al. 2015) or other plazas (Cap 2015; King 2015). Marketplace trade provided access by the broad spectrum of society to commodities from near and distant lands, including obsidian and Belize Red serving bowls at Caracol, for example (Chase et al. 2015). Bulk trade of corn and other foods at marketplaces, even beyond a city-state's region, may have balanced shortages and surpluses in the tropical rainforest setting where food storage is problematic (Eppich and Freidel 2015). Villagers from Ceren had choices of markets in various communities (Sheets et al. 2015), as did the Maya at larger communities and cities. Currency equivalencies, including cacao, marine shells, and standard sizes of cotton cloth, must have been used for the multiple exchanges that took place (Freidel et al. 2016; McAnany 2010:255). The Paynes Creek research indicates that salt cakes in standard sizes may also have been used as currency equivalencies.

Marketplace trade was one of several different mechanisms of exchange of commodities during the Classic period. Another network of exchange was based on the production of other resources on the coast, such as cacao in the Port Honduras region at the Village Farm site, that were transported inland and farther away along the coast. Cacao was likely added to canoes transporting salt cakes and salted fish up rivers in southern Belize. A third network of exchange focused on extraction of marine resources that were in demand by the inland Maya for ritual and other uses, such as stingray spines for bloodletting, conch shells for trumpets, shells for decoration, coral rock, pumice, and manatee bones. Another network of exchange was focused on coastal trading ports that brought commodities from farther away. In the case of the Paynes Creek Salt Works, commodities from more distant areas included obsidian, jadeite, and other greenstones.

Trading ports along the coasts of Belize and the Yucatán of Mexico facilitated the transportation of commodities along the coast. Cacao may have been transported along the coast to areas where cacao was in demand but not locally grown, as reported at the time of European contact in the sixteenth century. Commodities from northern Belize include chert with tool assemblages similar to those produced at Colha, which were excavated at Wild Cane Cay (McKillop 2005b), the Paynes Creek Salt Works, and the Village Farm site. Belize Red pottery made in the upper Belize Valley was found at the Paynes Creek Salt Works and the Village Farm site. The Altun Ha pottery vase from Copan may have been transported along the coast. Mercury from Honduras found under the central ball court marker at Lamanai (Pendergast 1982) was likely transported along the coast and up the New River. The pattern of obsidian source use at the Paynes Creek Salt Works matches the obsidian source use at Wild Cane Cay and contrasts to the pattern at inland communities, indicating there were different exchange networks along the coast and inland. Ixtepeque sourced obsidian was more common than El Chayal at the salt works and at Wild Cane Cay, whereas inland communities of Lubaantun, Uxbenka, and Pusilha had overwhelming abundance of El Chayal obsidian. On the coast, obsidian was traded to Wild Cane Cay where prismatic blades were struck for distribution to the salt works and other sites in the Port Honduras region that lack cores (McKillop 2005b). Jadeite and other greenstone objects recovered from the Paynes Creek Salt Works were from the Motagua River outcrops, either on the north or south shores (McKillop et al. 2019). The Motagua River is a short canoe paddle from Port Honduras, a voyage that may have been accomplished by Maya from Wild Cane Cay or by traders from the jadeite outcrops, where there are workshops for preliminary working (Rochette 2009).

Classic Maya salt production at the Paynes Creek Salt Works and elsewhere along the coast of Belize and the Yucatán supplied salt and salted fish to nearby inland communities, along with other commodities added to the loads. Bones from marine fish were recovered from Lubaantun (Wing 1975) and are commonly found at nearby inland communities such as Altun Ha and Dzibilchaltún, and even farther inland at Tikal (McKillop 1985, 2006, 2007a). High levels of strontium in human bone from burials at Tikal may indicate that Caribbean sea salt was imported (Fenner and Wright 2014; Wright 2005). Bulk canoe transport up rivers in Belize gave mariners control of distribution in addition to production. A lack of rivers in the Yucatán meant that overland porters following trails transported salt and other commodities but that the producers may not have been the traders or may have been part of contingent multicrafting. Additional commodities for the inland salt trade in southern Belize likely included cacao produced on the mainland coast of Port Honduras. The Village Farm site is the location of a modern organic cacao farm with over 5,000 cacao trees. There are additional rich agricultural soils between Deep River and Golden Stream in the Port Honduras region. The Village Farm site is contemporary with the Paynes Creek Salt Works and shares ceramic and lithic assemblages, notably ocarinas, unit-stamped pottery, and Colha chert.

Also added to the inland transport of salt were marine resources for ritual and other uses as well as commodities transported along the coast from farther away (McKillop 1996a, 2005b, 2016b; McKillop and Healy 1989). Stingray spines used in ritual bloodletting ceremonies were found at Nim Li Punit (Figure 8.4; Jordan and Prufer 2017). Large conch shells (*Turbinella angulata*) were used as musical instruments at inland communities and are depicted on the painted murals at Bonampak (Miller 1999). Coastal trading ports along the coast of Belize and the Yucatán provided safe harbor from weather and long canoe trips, places where goods transported along the coast were added to inland trade, and even warehouses as on Cozumel island (Freidel and Sabloff 1984). Wild Cane Cay was ideally located for coastal trade and for integrating the coastal and inland transportation networks of commodities. With its location at the north end of two ranges of cays in Port Honduras, Wild Cane Cay was the last harbor for coastal canoe travelers and traders heading farther north along the coast of Belize. The protected harbor provided anchorage for boats and cargo as well as hospitality at the 10-acre village. Until the southern highway was built in the 1960s, ferries transporting mail, cargo, politicians, priests, and other travelers stopped at Wild Cane Cay. Other trading ports along the coasts of Belize and the Yucatán peninsula of Mexico include False Cay by Placencia; Moho Cay in the mouth of the Belize River (McKillop 2004); Saktunja and Northern River Lagoon on the north coast of Belize (Masson and Mock 2004); Marco Gonzalez (Graham 1989) and San Juan (Guderjan and Garber 1995) on Ambergris Caye; Cerros (Robertson 2016; Walker 2016) and Santa Rita at Corozal (Chase 1981) on Chetumal Bay; Cozumel Island (Sabloff and Rathje 1975); Isla Cerritos (Andrews et al. 1989); and Xcambó (Sierra Sosa et al. 2014). The Terminal Classic rise of Chichén Itzá is attributed to the expansion of

Gulf of Mexico Chontalpa traders who extended their interest in controlling salt to the north coast salt works and then Chichén Itzá, bringing a new style of architecture (virtually identical to some buildings at Tula, in central Mexico), and gaining control of sea trade (Andrews and Robles 1985). Distribution of round structures and imported slate ware pottery extended to Belize (Harrison-Buck and McAnany 2013).

Salt-Related Activities

Activities contingent to salt production at the Paynes Creek Salt Works included extracting and transporting lumber, wood fuel, and palm thatch by canoe from several nearby ecosystems for construction and maintenance of the salt kitchens. Clay and quartz sand were obtained nearby on cut banks along the pine savanna and transported by canoe for pottery production. The pottery paddle from Site 103 may indicate one such locale. Leaching brine through salty soil probably occurred at each salt kitchen, a necessary step prior to evaporation. Evaporation ponds may also have been used, as suggested by drone imagery.

Some of the contingent activities, including transporting salt cakes, salted fish, and other commodities to market, may have been outsourced (Arnold 2017). Commercial fishing takes an investment of time and expertise as well as good fishing grounds. Punta Ycacos Lagoon is one of the best fly-fishing areas in the world, known for permit and bonefish, but there also are goliath groupers, barracudas, manatees, crocodiles, jaguars, peccary, frigate birds, and brown pelicans. Use-wear analysis indicates about 80 percent of the chert stone tools examined by Kazuo Aoyama were used exclusively for cutting fish or meat, which means a tremendous amount of salt was produced for salting fish or meat (McKillop and Aoyama 2018). That the 4,042 posts and other architectural wood mapped had chipped ends, that about 20 percent of the chert tools had use-wear indicating cutting or whittling wood, and that recovery of a sample of what must have been a sizable woodworking industry for canoes, paddles, pottery paddles, and other items for use in the salt kitchens, fishing and processing fish, and household needs, means that woodworking required time and skill, and some of it may have been specialized or outsourced. The salt workers needed food, whether they lived in residences near their salt kitchens or at Wild Cane Cay, Village Farm, or another nearby community. Fish, a variety of tree crops that may have grown by the salt kitchens when sea level was lower and the land was dry, and oysters growing on the red mangrove prop roots all required collection, storage, and preparation. Bulk corn and other food commodities provided ballast on the return trips from inland marketplaces as well as carbohydrates to add to the coastal diet. Water jars for storing brine, loose salt, and drinking water were obtained from inland markets. Drinking water was obtained from nearby wells and seeps, up rivers, and from nearby fresh water ponds.

The preservation of wood at the Paynes Creek Salt Works indicates there was a substantial infrastructure of production and distribution of salt, which provides a model for the economics of salt production along the coast of Belize and the Yucatán of Mexico, where wood has not preserved. Mapping artifacts on the sea floor indicates that salt production was carried out indoors, similar to the salt kitchens at Sacapulas in the Maya Highlands and that the residences were either in other buildings at the salt works, at Wild Cane Cay, or at another nearby coastal location. The infrastructure of distribution consists of canoe transportation documented by the first known ancient Maya canoe paddle and canoe as well as other canoe paddle blade fragments. Use-wear on chert tools and the chipped ends of 4,042 wooden posts and other architectural elements further underscores that the woodworking industry was contingent upon multicrafting.

Surplus household production of salt, at times geographically distant from residences, was carried out along the coast of Belize and the Yucatán. Dynastic Maya control of production was limited to management and perhaps fees at marketplaces (Chase et al. 2015; Jones 2015). Elite management of salt production may have occurred at the large inland salt-production facility at Salinas de los Nueve Cerros, suggested because of the elite architecture beside the salt works (Woodfill et al. 2015). If this were the case, it presaged the apparent elite control of salt production at Emal in the Postclassic (Kepecs 2003a) and perhaps also at Xcambó during the Classic period (Sierra Sosa et al. 2014). In China during the dynastic era, salt was considered an important commodity that was monopolized by the state and played an important role in state finance (Flad 2011, 2013; Fu 2015; Li and Falkenhausen 2006, 2010). Worldwide, the organization of salt production and distribution of rock salt, solar evaporation at salt flats, and evaporation by heating brine in pots over fires was focused on the human need for salt, both for the biological and additive qualities of salt. At times the state intervened at production facilities by placing state overseers, as at the salt works in Shandong Province in southeast China (Flad 2013). The state sometimes intervened in distribution, as with the Inca control and distribution of commodities throughout the empire or the Romans doling out salt as salary to soldiers. And the state sometimes taxed salt and other commodities at the marketplace, as in the sixteenth-century markets in the Yucatán (Andrews 1983; King 2015:59) and as indicated by marketplaces and association with elite architecture among the Classic Maya at Caracol, Calakmul, and Tikal. As a scarce resource with restricted natural distribution, the value of salt varied, sometimes worth its weight in gold, as with the caravans of camels loaded with salt from mines at Taoundenni taking their cargo across the Sahara to Timbuktu (Benanav 2006).

The Economics of Salt Production and Distribution

Evaluating the scale, concentration, intensity, context, and constitution of labor (Costin 1991) at the Paynes Creek Salt Works provides a framework for evaluating the economics of salt production and distribution.

Scale

The salt works are a single, large salt-production complex with salt kitchens and perhaps other wooden buildings spread out over about 10 km^2, with most buildings concentrated within a 5 km^2 area. During survey, individual sites were identified by clusters of artifacts and wooden posts visible on the sea floor that were separated by open space from other clusters. In order to address the scale of production, I needed to evaluate if all the salt works were used at one time, indicating a massive effort at producing salt or, instead, if a few localities were used at a time over several hundred years, indicating small-scale production. Radiocarbon and accelerator mass spectrometry dating of wooden posts provided a time range from the late Early Classic into the Early Postclassic (AD 400–1100), with much overlap in date ranges during the Late Classic. Modal analyses of nonbriquetage pottery places most of the sites in the Late and Terminal Classic, contemporary with the population rise at consumer populations at inland communities in southern Belize and generally in the Southern Maya Lowlands.

Examination of salt kitchens along an ancient shoreline in the East

Lagoon was carried out to evaluate if the salt kitchens were constructed and used at the same time, indicating large-scale production, or if they were built one or more at a time with later infilling. The results indicate that adjacent Sites 24 and 25/26 have similar patterns of wood selection for building posts, with radiocarbon dates within the same ranges. By way of contrast, Site 23 has a different pattern of wood selection and a later range on the radiocarbon date from a post at the site. The findings support the interpretation that a few salt kitchens were constructed along shorelines with later infilling. When sea-level rise breached the seawall protection of the lines of palmetto palm posts on the edges of the yards of the salt kitchens and no longer kept water from rushing into the yards of the salt kitchens, the salt kitchens were moved inland. The process continued, resulting in several lines of salt kitchens in the East Lagoon and the West Lagoon.

Concentration (Dispersed to Nucleated)

I used standardization of briquetage and tree species identification of wooden posts along with the radiocarbon and accelerator mass spectrometry dates on wooden posts to suggest that the Paynes Creek Salt Works were used by separate work parties that were not all used at the same time. Briquetage from three salt works indicated that the pots were standardized in dimensions in contrast to household pottery from nearby Wild Cane Cay and that the values were statistically significant. In addition, there were statistically significant differences in standardization among three salt works, the Stingray Lagoon site, David Westby, and Orlando's, indicating the salt kitchens at each site were used by different work parties, likely households. The salt pots were standardized to produce standard units of salt in the form of hardened salt cakes or hardened salt stored and transported in pots, as described for Bohol Island in the Philippines (Yankowski 2010). Standard units of salt facilitated marketplace trade in equivalencies as with units of cloth and cacao (Baron 2018a, 2018b). Tree species identification of wooden posts revealed similar patterns of wood selection at David Westby and Orlando's, whereas Stingray Lagoon site was different and in fact consisted of two different construction efforts.

Intensity

The range of activities directly related to salt production and other supporting and related activities such as salting fish and woodworking, for example, indicate considerable time and effort best explained by surplus household production, with contingent multicrafting and likely outsourcing of some contingent activities to extended family members or others.

Context (Independent to Attached Specialists)

Salt making at the Paynes Creek Salt Works developed from surplus household production involving many related activities that were carried out year-round by coastal Maya who lived nearby on the coast, including the possibility of residences at the salt works. Wooden buildings at larger sites, notably Site 7 and Ek Way Nal, may have included residences. The spatial patterning of artifacts at Site 7 revealed that briquetage was associated with some but not all of the wooden buildings. In addition, a variety of domestic activities are indicated by the recovery of a spindle whorl, glyph vase, and snuff bottle, among others at Site 7. Site 7 may have been used later than Ek Way Nal, which has Late Classic ocarinas and unit-stamped jar motifs. Wild Cane Cay and the Village Farm site are also candidates for residences.

Constitution of Labor (Kin-Based to Industrial)

The constitution of labor has already been discussed as consisting of surplus household production.

Salt Yields in the Late Classic Maya Area

Salt works on the coast of Belize may have produced enough salt for the Late Classic salt needs in the Southern Maya Lowlands, along with contributions from Salinas de los Nueve Cerros, Maya Highland salt springs, and the north coast of the Yucatán. Coastal salt production in Mexico, Belize, and Guatemala supplied nearby inland Maya communities, in the Northern Maya Lowlands, Southern Maya Lowlands, and near the coast in the Southern Maya Highlands.

Discovery of salt works using the briquetage technique along the coast of Belize provides missing data for salt production and supply to the inland Maya in the Southern Maya Lowlands. The lack of known salt works along the coast of Belize previously supported a model of importing north coast salt to the Southern Maya Lowlands (Andrews 1983:Figure 7.1, Table 7.2). Even after the discovery of Belizean salt works—many submerged and hidden underwater—a view that north coast salt was superior in quality to Belize salt supported a model of import of high-quality salt and local use of low-quality Belize salt (MacKinnon and Kepecs 1989; Andrews and Mock 2002). The discovery and excavation of salt works along the coast and on offshore islands in Belize indicates that sufficient salt was produced to satisfy the dietary needs of nearby inland Maya during the Late Classic and that northern Yucatecan coast salt was not imported at that time. Even after the clear incursion of northern coast maritime traders with their base at Chichén Itzá during the Terminal Classic, the Belize salt works were still very active, and cities farther inland were being abandoned.

The main salt works along the Belize coast dated to the Late and Terminal Classic (with some dating as early as the Late Preclassic) include Northern River Lagoon site, 11 sites on Northern River Lagoon, Saktunja, 12 sites on Midwinter's Lagoon, Moho Cay, Wits Cah Ak'al, Marco Gonzalez and other sites on Ambergris Caye, Placencia, and the Paynes Creek Salt Works. Using estimates from Sacapulas, where two people making salt in a salt kitchen could produce 6 tons of salt per year (Reina and Monaghan 1981), an estimated 60–600 tons of salt per year was made at the 10–100 salt kitchens at the Paynes Creek Salt Works. Adding the same yields for eight additional Belize salt works (not including all salt works), results in 480 to 4,800 tons of salt per year produced along the coast of Belize during the Late and Terminal Classic. With Anthony Andrews's (1983:114, Table 7.1) estimate of 5 million people in the Late Classic who needed 15,000 tons of salt per year, the salt needs of the Maya were met by the Belize coastal salt works (480–4,800 tons per year), Salinas de los Nueve Cerros (300 tons per year estimated by Andrews 1983—which is probably a low estimate based on field research by Woodfill et al. 2015), and the north coast salt works (20,000 tons). Adding Sacapulas (6 tons × 50 salt kitchens = 300 tons), San Mateo Ixtatán (300 tons), and other salt works in the highlands brings the total annual production to at least 26,500 tons of salt. Andrews (1983:138) estimates north coast salt production was about 20,000 tons in early colonial times before modern mechanized salt production. Additional Maya Highland salt works may also have contributed to Maya Lowland salt needs. Andrews (1983:60–61) estimates production of about 35 tons of salt a year at Ixtápa and 60–100 tons of salt if all seven salt kitchens were used all years, producing enough salt for 35,000 people. Even with population estimates of 7–11 million people in the central Maya lowlands based on recent LiDAR mapping (Canuto et al. 2018), salt production yields of 26,500 tons per year exceeded dietary needs for 8 million people.

As a comparison, estimates of an annual demand for salt in the Roman Empire are 350,000 tons to meet the needs of a population of over 50 million people (Saile 2015). This calculation includes the salt needs for livestock and food preservation, based on a yearly per capita need of 30 kg (of which 7 kg per capita annually for human dietary salt needs alone). The Romans expanded coastal salt production by solar evaporation along the Mediterranean Sea and coastal areas of northwest France while prohibiting trade from inland briquetage and rock salt works in the interior of France, Germany, and Austria (Saile 2015).

Summary

The Paynes Creek salt workers were specialists who produced salt by boiling brine in salt kitchens as part of surplus household production. They were artisanal specialists who controlled the flavor, texture, and color of the salt they produced. Fine white table salt composed of sodium chloride was not a desired outcome. The salt workers adjusted the length and intensity of boiling, the soil or sediment selected for enriching the brine before it was boiled, and whether fresh water or sea water was filtered through the soil in the enrichment process. The color was also adjusted when salt cakes were hardened on a fire, following the initial evaporation process. These factors altered the chemical composition of the salt. By controlling the length and intensity of evaporation, the salt workers—knowingly or not—controlled the chemical composition of the salt: Different salts precipitate out of brine at different rates. Calcium salts precipitate first, followed by sodium chloride, then magnesium sulfate and potassium salts, and finally magnesium chlorine (bittern). A low heat with a long evaporation is used by Japanese salt makers to ensure the salt includes bittern, which they consider tasty (Bitterman 2010:39). Jeffrey

Parsons (2001) notes that salt makers in central Mexico selected different soils for different flavors of salt. In some cases, a generic salt was desired, regardless of the chemical composition or flavor (Gouletquer and Weller 2015:4).

Various tasks were necessary for salt production, best described as contingent multicrafting. Some of the contingent activities were carried out by the householders, but other tasks may have been outsourced to relatives or even to other specialists. The bowls, jars, and basins used to boil the brine may have been made by local potters who obtained the clay and quartz sand for temper near the salt works and made coiled pots using pottery paddles similar to the one from Site 103. Perhaps that site included a potter's workshop. Transporting the salt cakes, salted fish, and other marine products by boat up rivers to marketplaces may have been outsourced since the travel would have reduced time available for salt production.

Salt was produced for household and local community needs, but the bulk of salt was produced as salt cakes and used to make salted fish for marketplace trade. Standard sizes of salt cakes were made either by using forms or by hardening the salt in pots. The large quantities of briquetage in excavations indicates at least some brine-boiling vessels were discarded at the salt kitchens. For example, 98 percent of the ceramics from excavations in Building A at Site 74 consisted of briquetage. Although salt cakes would not leave physical remains of vessels at marketplaces or the in the homes of inland consumers, salt cakes traded inside pottery vessels would leave Punta Ycacos Unslipped pots at the inland communities. If some of the base was broken to expose the salt cake inside the pot, then base sherds would be found at the salt kitchens or nearby discard areas but not the rest of the pots. Salt cakes may have been wrapped in leaves, perhaps in pots, as on Bohol Island in the Philippines. In that scenario, organic residue and anything else adhering to the leaves would be delivered to inland consumers.

The inland consumers of salt cakes and salted fish left their identity at the salt works by the diverse origins of unit-stamped pottery. Compositional analyses indicate there were at least 13 sources for the unit-stamped pots. Other nonlocal pottery such as Belize Red serving dishes and figurine whistles may have been obtained by the salt traders at inland marketplaces.

The residences of the salt workers were either at the salt works or nearby on the coast or cays. Seasonality of tree fruits, including mamey apple, craboo, coyol, and cohune that were commonly recovered from sea floor survey and from excavations, indicate year-round use of the salt works. The tree fruits were radiocarbon dated to the Classic period from sea floor survey of Site 56 and were abundant in excavations inside Building A at Site 74. With upward of 10 buildings at Sites 7 and 60 and perhaps elsewhere such as Sites 72 and 58, some of the buildings may have been residences while others were salt kitchens or were used for other related tasks.

The use of salt kitchens meant that production could be year-round, even in the rainy season. The high season for salt production was April to mid-June when rainfall is minimal. Brine-enrichment could occur year-round during breaks in the rainfall, such as the dry weather in July and August. Storing brine in jars and salty soil inside salt kitchens as at Sacapulas meant that salt could be made at any time of year, even in persistent rain. Salt could also be stored indoors and transported to market year-round. Lines of sites in the East and West Lagoons may represent sequential salt kitchens being moved as a result of rising seas. Alternatively, residences may have been located behind the salt kitchens as at Sacapulas and in southeast China in Shandong Province. Some or all of the salt workers may have traveled from their homes on Wild Cane Cay or Village Farm, bringing fish to salt near the salt kitchens. Although fish are plentiful in the lagoon, there is more diversity in the open water, including the area around Wild Cane Cay. The fish species identified at Lubaantun, one of the consumer communities for Paynes Creek salt and salted fish, are all found in the waters around Wild Cane Cay, and their bones were excavations in Late to Terminal Classic middens at the site.

The Paynes Creek Salt Works developed from surplus household production, with contingent multicrafting (De Lucia 2013; Hirth 1998) that was integrated into the Classic Maya economy through marketplace trade. The salt workers were participants in various networks of exchange involving near and distant commodities as well as marketplace trade and reciprocity (see also Demarest et al. 2014). The salt workers transported salt, salted fish, marine resources, and cacao to inland markets and returned with corn, pots, and ocarinas. Professional Maya traders, *p'olom* (King 2015:46), such as those observed at European contact in the sixteenth century, likely participated in longer trading trips that integrated coastal trading ports such as Wild Cane Cay into a circum-Yucatán canoe

Figure 8.2. Exhibit featuring 3D-printed replicas of 3D-scanned artifacts and wooden posts from the Paynes Creek Salt Works, at the Tourism Information Center, Punta Gorda, hosted by Heather McKillop and E. Cory Sills, "Sustainable Archaeological Tourism of the Paynes Creek Underwater Maya Project."

trade system. God L of the Classic period, who became known as Ek Chuah in the Postclassic, was the patron of merchants and cacao (McAnany 2010:255–257).

Salt was certainly one of the movers and shakers in Classic Maya society. Many of the waterlogged artifacts from the Paynes Creek Salt Works have been 3D scanned to preserve a digital research record of the ancient salt industry (McKillop and Sills 2013a, 2013b). The use of 3D-printed replicas of the artifacts in exhibits, lectures, and public outreach provides a tangible way of sharing the cultural heritage (Figures 8.2–8.3). Submerged by sea-level rise, the Paynes Creek Salt Works are a sobering reminder of low-lying cities and other communities worldwide subject to rising seas from global warming.

Figure 8.3. "Museum in a Bag," traveling display of 3D-printed replicas of artifacts from the Paynes Creek Salt Works with a backdrop of salt crystals.

NOTES

Chapter 1. Economics of Maya Salt

1. Although some researchers argue that the Late Classic Maya were centrally organized (Chase and Chase 1996), others suggest that their political organization was decentralized, or that it may have fluctuated over time or geographically between centralized and decentralized modes (Marcus 1993). The Classic Maya city of Caracol was centrally controlled by the dynastic leaders, as shown by causeways linking remote administrative nodes to the epicenter (Chase and Chase 2014).

Chapter 4. The Infrastructure of Salt Production: Wooden Architecture

1. Summarized from daily rainfall data collected by Tanya Russ and John Spang at Village Farm, located on the coast between Middle River and Seven Hills Creek, in the Port Honduras Marine Reserve.

2. In the GIS Geomedia, I used the thematic category under Legends to classify the attribute "diameter" as a range thematic for size categories. I varied the ranges from equal numbers to selected ranges, including selecting the size ranges for building posts reported by Robert Wauchope (1938) in his study of modern Maya houses (0–8cm; 8–16 cm; 16–24cm; 24–32 cm; unknown).

Chapter 5. Wood Selection for the Salt Kitchens

1. Species identification of wood posts was begun in 2009 by project wood anatomist Mike Wiemann and continued as a PhD dissertation with a case study of Early Classic Site 24 and Late Classic Site 35 by Mark Robinson (Robinson and McKillop 2013) and for other sites as part of his work in the lab. Wiemann supervised Robinson for three months at the USDA Wood Products Lab in identifications, acquiring knowledge, which he shared on return to LSU.

Chapter 6. Labor, Production, and Identity

1. Wooden objects, including the K'ak' Naab' canoe paddle, were conserved using the polymer process (Smith 2003) at the Texas A&M Preservation Lab by C. Wayne Smith and Helen Dewolf. The polymer process allows wood and other objects to be without humidity or temperature control, which is useful for artifacts returned to Belize.

2. Radiocarbon dates are reported with their 2 Sigma range, indicating that there is a 95 percent likelihood that the age of the sample is somewhere within the range.

3. Divisions within the Early Classic (Tzakol 1–3), Late Classic (Tepeu 1–3), and Terminal Classic (Tepeu 3) follow Gifford 1976.

4. In Thailand, women form work parties for salt production (Halliwell et al. 2016; Yankowski et al. 2015).

REFERENCES CITED

Abrams, Elliot M.

1994 *How the Maya Built Their World: Energetics and Ancient Architecture*. University of Texas Press, Austin.

Adams, Richard E. W.

1971 *The Ceramics of Altar de Sacrificios*. Papers of the Peabody Museum of Archaeology and Ethnology 63(1). Harvard University, Cambridge, Massachusetts.

Adshead, S.A.M.

1992 *Salt and Civilization*. St. Martin's, New York.

Aikman-Smith, Valerie

2009 *Salt: Cooking with the World's Favorite Seasoning*. Ryland, Peters, and Small, New York.

Aimers, James, Elizabeth Haussner, Dorothy Farthing, and Saturo Murata

2016 An Expedient Pottery Technology and Its Implication for Ancient Maya Trade and Interaction. In *Perspectives on the Ancient Maya of Chetumal Bay*, edited by Debra S. Walker, pp. 149–161. University Press of Florida, Gainesville.

Akridge, D. G.

2008 Methods for Calculating Brine Evaporation Rates during Salt Production. *Journal of Archaeological Science* 35:1453–1462. doi.org/10.1016/j.jas.2007.10.013.

Andrews, Anthony P.

1983 *Maya Salt Production and Trade*. University of Arizona Press, Tucson.

Andrews, Anthony P., Frank Asaro, Helen V. Michel, Fred H. Stross, and Pura Cervera Rivero

1989 The Obsidian Trade at Isla Cerritos, Yucatan, Mexico. *Journal of Field Archaeology* 16:355–363. doi.org/10.1179/jfa.1989.16.3.355.

Andrews, Anthony P., and Shirley B. Mock

2002 New Perspectives on the Prehispanic Maya Salt Trade. In *Ancient Maya Political Economies*, edited by Marilyn A. Masson and David A. Freidel, pp. 307–334. Altamira, New York.

Andrews, Anthony P., and Fernando Robles C.

1985 Chichén Itzá and Cobá. In *The Lowland Maya Postclassic*, edited by Arlen F. Chase and Prudence M. Rice, pp. 62–72. University of Texas Press, Austin.

Andrieu, Chloé

2013 Late Classic Maya Lithic Production and Exchange at Rio Bec and Calakmul,

Mexico. *Journal of Field Archaeology* 38:21–37. doi.org/10.1179/009346901 2Z.00000000034.

Antonites, Alexander

2016 The Organization of Salt Production in Early First Millennium CE South Africa. *Journal of Anthropological Archaeology* 44:31–42. doi.org/10.1016/j .jaa.2016.08.001.

Aoyama, Kazuo

2017 Ancient Maya Economy: Lithic Production and Exchange around Ceibal, Guatemala. *Ancient Mesoamerica* 28:279–303. doi.org/10.1017/S0956536116000183.

Ardren, Traci, Alejandra Alonso Olvera, and T. Kam Manahan

2016 The Artisans of Terminal Classic Xuenkal, Yucatan, Mexico: Gender and Craft during a Time of Economic Change. In *Gendered Labor in Specialized Economies*, edited by Sophia K. Kelly and Traci Ardren, pp. 91–115. University Press of Colorado, Boulder.

Arnold, Dean

2017 Changes in Ceramics as Commodities in Ticul, Yucatán, Mexico (1965–2008) and What They Tell Us about Ancient Maya Ceramic Production. In *The Value of Things: Prehistoric to Contemporary Commodities in the Maya Region*, edited by Jennifer P. Matthews and Thomas H. Guderjan, pp. 193–214. University of Arizona Press, Tucson.

Awe, Jaime, and Paul F. Healy

1994 Flakes to Blades? Middle Formative Development of Obsidian Artifacts in the Upper Belize River Valley. *Latin American Antiquity* 5:193–205. doi.org/10.2307/971879.

Ball, Joseph W., and Jennifer T. Taschek

2001 The Buenavista-Cahal Pech Royal Court: Multi-Palace Court Mobility and Usage in a Pottery Lowland Maya Kingdom. In *Royal Courts of the Maya*, vol. 2, *Dates and Case Studies*, edited by Takeshi Inomata and Stephen Houston, pp. 165–200. Westview, Boulder, Colorado.

Baron, Joanne

2018a Ancient Monetization: The Case of Classic Maya Textiles. *Journal of Anthropological Archaeology* 49:100–113. doi.org/10.1016/j.jaa.2017.12.002.

2018b Making Money in Mesoamerica: Currency Production and Procurement in the Classic Maya Financial System. *Economic Anthropology* 5(2):210–223. doi:10.1002/sea2.12118.

Benanav, Michael

2006 *Men of Salt: Crossing the Sahara on the Caravan of White Gold*. Lyons, Guilford, Connecticut.

Bill, Cassandra, and Geoffrey Braswell

2005 Life at the Crossroads: New Data from Pusilha, Belize. *Research Reports in Belizean Archaeology* 2:301–312.

Bishop, William K.

1977 Coastal Salt-Making along the Gulf of Nicoya, Costa Rica: A Historical-Cultural Geography. M.A. thesis, Department of Geography and Anthropology, Louisiana State University, Baton Rouge.

Bitterman, Mark

2010 *Salted: A Manifesto on the World's Most Essential Mineral, with Recipes*. Random House, New York.

Braswell, Geoffrey E.

2010 The Rise and Fall of Market Exchange: A Dynamic Approach to Ancient Maya Economy. In *Archaeological Approaches to Market Exchange in Ancient Societies*, edited by Christopher P. Garraty and Barbara L. Stark, 127–140. University Press of Colorado, Boulder.

Braswell, Geoffrey, Cassandra Bill, and Christian Prager

2007 Exchange, Political Relations, and Regional Interaction: The Ancient City of Pusilha in the Late Classic Maya World. *Research Reports in Belizean Archaeology* 5:51–62.

Braswell, Geoffrey, Christian M. Prager, Cassandra Bill, and Sonja Schwake

2004 The Rise of Secondary States in the Southeastern Periphery of the Maya World: A Report on Recent Archaeological and Epigraphic Research at Pusilha, Belize. *Ancient Mesoamerica* 15:219–233. doi.org/10.1017/S0956536104040143.

Braud, Melissa R.

1996 Underwater Test Excavations at David Westby Site, South Coastal Belize. M.A. thesis in Anthropology, Department of Geography and Anthropology, Louisiana State University, Baton Rouge.

Brigand, Robin, and Olivier Weller (editors)

2015 *Archaeology of Salt: Approaching an Invisible Past*. Sidestone, Leiden.

Brown, Ian

1980 *Salt and the Eastern North American Indian: An Archaeological Study*. Lower Mississippi Survey, Bulletin No. 6. Peabody Museum, Harvard University, Cambridge, Massachusetts.

Canuto, Marcello A., Fransisco Estrada-Belli, Thomas G. Garrison, Stephen D. Houston, Mary Jane Acuna, Milan Kovac, Damien Marken, Philippe Nondedeo, Luke Auld-Thomas, Cyril Castanet, David Chatelain, Carlos R. Chiriboga, Tomas Drapela, Tibor Lieskovsky, Alexandre Tokovinine, Antolin Velasquez, Juan C. Fernandez-Diaz, and Ramesh Shrestha

2018 Ancient Lowland Maya Complexity as Revealed by Airborne Laser Scanning of Northern Guatemala. *Science* 361(6409). doi.org/10.1126/science.aau0137.

Cap, Bernadette

2015 How I Know It When I See It: Marketplace Identification at the Classic Maya Site of Buenavista del Cayo, Belize. In *The Ancient Maya Marketplace: The Archaeology of Transient Space*, edited by Eleanor King, pp. 111–137. University of Arizona Press, Tucson.

Carrasco, Ramón Vargas, Verónica A. Vázquez López, and Simon Martin

2009 Daily Life of the Ancient Maya Recorded on Murals at Calakmul, Mexico. *Proceedings of the National Academy of Sciences* 106:19245–19249. doi.org/10.1073/pnas.0904374106.

Chase, Arlen F., and Diane Z. Chase
1996 More than Kin and King: Centralized Political Organization among the Late Classic Maya. *Current Anthropology* 37:803–810.
2012 Belize Red Ceramics and Their Implications for Trade and Exchange in the Eastern Maya Lowlands. *Research Reports in Belizean Archaeology* 9:3–14.
Chase, Arlen F., Diane Z. Chase, Richard E. Terry, Jacob M. Horlacher, and Adrian S. Z. Chase
2015 Markets among the Ancient Maya: The Case of Caracol, Belize. In *The Ancient Maya Marketplace: The Archaeology of Transient Space,* edited by Eleanor King, pp. 226–250. University of Arizona Press, Tucson.
Chase, Arlen F., and Prudence M. Rice (editors)
1985 *The Lowland Maya Postclassic.* University of Texas Press, Austin.
Chase, Diane Z.
1981 The Maya Postclassic at Santa Rita Corozal. *Archaeology* 34(1):25–33.
Chase, Diane Z., and Arlen F. Chase
2014 Ancient Maya Markets and the Economic Integration of Caracol, Belize. *Ancient Mesoamerica* 25:239–250. doi.org/10.1017/S0956536114000145.
Christie, Jessica J. (editor)
2003 *Maya Palaces and Elite Residences: An Interdisciplinary Approach.* University of Texas Press, Austin.
Clark, John E.
1986 From Mountains to Molehills: A Critical Review of Teotihuacan's Obsidian Industry. In *Research in Economic Anthropology*, Supplement 2, edited by Barry L. Isaac, 23–74. JAI, London.
Coe, Michael D., and Kent V. Flannery
1967 *Early Cultures and Human Ecology in South Coastal Guatemala.* Smithsonian Contributions to Anthropology 3. Smithsonian Institution, Washington DC. doi.org/10.5479/si.00810223.3.1.
Coe, Michael, and Justin Kerr
1998 *The Art of the Maya Scribe.* Harry N. Abrams, New York.
Coggins, Clemency, and Orrin C. Shane III (editors)
1984 *Cenote of Sacrifice: Maya Treasures from the Sacred Well at Chichén Itzá.* University of Texas Press, Austin.
Coles, B., and J. M. Coles
1989 *People of the Wetlands.* Guild, New York.
Coronel, Eric G., Scott Hutson, Aline Magnoni, Chris Balzotti, Austin Ulmer, and Richard Terry
2015 Geochemical Analysis of Late Classic and Postclassic Maya Marketplace Activities at the Plazas of Cobá, Mexico. *Journal of Field Archaeology* 40:89–109. doi.org/10.1179/0093469014Z.000000000107.
Costin, Cathy L.
1991 Craft Specialization: Issues in Defining, Documenting, and Explaining the Organization of Production. In *Archaeological Method and Theory*, Vol. 1, edited by Michael B. Schiffer, 1–56. University of Arizona Press, Tucson.

Croes, Dale (editor)

1976 *The Excavation of Water-Saturated Archaeological Sites (Wet Sites) on the Northwest Coast of North America*. National Museum of Canada, Ottawa.

Culbert, T. Patrick, and Don S. Rice (editors)

1990 *Precolumbian Population History in the Maya Lowlands*. University of New Mexico Press, Albuquerque.

Curtis, Robert I.

1984 The Salted Fish Industry of Pompeii. *Archaeology* 37:58–59, 74–75.

Dahlin, Bruce H., Christopher T. Jensen, Richard E. Terry, David R. Wright, and Timothy Beach

2007 In Search of an Ancient Maya Market. *Latin American Antiquity* 18:363–384. doi.org/10.2307/25478193.

Daltabuit, Magali, Alicia Rios, and Fraterna Perez

1988 *Coba: Estrategias Adaptativas de Tres Familias Mayas*. Instituto de Investigaciones Antropólogicas, UNAM Centro de Investigaciones de Quintana Roo, Serie Antropológica 90, Universidad Nacional Autónoma de México, Mexico City.

Daniels, James T., and Geoffrey E. Braswell

2014 Procurement, Production, and Distribution of Obsidian in the Southern Belize Region. *Research Reports in Belizean Archaeology* 11:289–296.

de Brisay, K. W., and K. A. Evans (editors)

1975 *Salt: The Study of an Ancient Industry*. Colchester Archaeological Group, Colchester.

De Leon, Jason P.

2009 Rethinking the Organization of Aztec Salt Production: A Domestic Perspective. In *Housework: Craft Production and Domestic Economy in Ancient Mesoamerica*, edited by Kenneth G. Hirth, pp. 45–57. Archeological Papers of the American Anthropological Association 19. doi.org/10.1111/j.1551-8248.2009.01012.x.

DeLeonardis, Lisa

2011 Itinerant Experts, Alternative Harvests: Kamajuq in the Service of Qhapaq and Crown. *Ethnohistory* 58:445–489. doi.org/10.1215/00141801-1263857.

De Lucia, Kristin

2013 Domestic Economies and Regional Transition: Household Multicrafting and Lake Exploitation in Pre-Aztec Central Mexico. *Journal of Anthropological Archaeology* 32:353–367. doi.org/10.1016/j.jaa.2013.05.001.

Demarest, Arthur

2004 After the Maelstrom: Collapse of the Classic Maya Kingdoms and the Terminal Classic in Western Peten. In *The Terminal Classic in the Maya Lowlands: Collapse, Transition, and Transformation*, edited by Arthur Demarest, Prudence M. Rice, and Don S. Rice, pp. 102–124. University of Colorado Press, Boulder.

Demarest, Arthur, Chloé Andrieu, Paola Torres, Mélanie Forné, Tomás Barrientos, and Marc Wolf

2014 Economy, Exchange, and Power: New Evidence from the Late Classic Maya Port City of Cancuen. *Ancient Mesoamerica* 25:187–219.

Dillon, Brian

1977 *Salinas de los Nueve Cerros, Alta Verapaz, Guatemala.* Ballena Press Studies in Mesoamerican Art, Archaeology and Ethnohistory 2. Ballena Press, Socorro, New Mexico.

Dillon, Brian, Kevin Pope, and Michael Love

1988 An Ancient Extractive Industry: Maya Saltmaking at Salinas de las Nueve Cerros, Guatemala. *Journal of New World Archaeology* 7(2/3):37–107.

Doran, Glen

2002 *Windover: Multidisciplinary Investigations of an Early Archaic Florida Cemetery.* University Press of Florida, Gainesville.

Dumas, Ashley A.

2007 The Role of Salt in the Late Woodland to Early Mississippian Transition in Southwest Alabama. PhD dissertation, Department of Anthropology, University of Alabama, Tuscaloosa.

Early, Ann M. (editor)

1993 *Caddoan Saltmakers in the Ouachita Valley: The Hardman Site.* Arkansas Archeological Survey Research Series No. 43. Arkansas Archeological Survey, Fayetteville.

Eaton, Jack D.

1978 Archaeological Survey of the Yucatán-Campeche Coast. In Studies in the Archaeology of Coastal Yucatán and Campeche, Mexico. Middle American Research Institute Publication 46(1). Tulane University, New Orleans.

Eppich, Keith, and David Freidel

2015 Markets and Marketing in the Classic Maya Lowlands: A Case Study from El Perú-Waká. In *The Ancient Maya Marketplace: The Archaeology of Transient Space*, edited by Eleanor King, pp. 195–225. University of Arizona Press, Tucson.

Eubanks, Paul N.

2013 Mississippian Salt Production at the Stimpson Site (1Ck29) in Southern Alabama. *Journal of Alabama Archaeology* 59:1–21.

Falkenhausen, Lothar von

2006 Background and Purpose of the Project. In *Ancient Salt Production and Landscape Archaeology in the Upper Yangzi Basin: Preliminary Studies. Salt Archaeology in China*, vol. 1, edited by Shuicheng Li and Lothar von Falkenhausen, pp. 11–27. Science Publishing, Beijing.

Falkenhausen, Lothar von, Li Shuicheng, and Sun Hua

2006 Conclusions and Prospects. In *Ancient Salt Production and Landscape Archaeology in the Upper Yangzi Basin: Preliminary Studies. Salt Archaeology in China*, vol. 1, edited by Shuicheng Li and Lothar von Falkenhausen, pp. 319–339. Science Publishing, Beijing.

Fash, William

2001 *Scribes, Warriors, and Kings: The City of Copan and the Ancient Maya.* Thames and Hudson, New York.

Feathers, Valerie, and Heather McKillop
2018 Assessment of the Shell Midden at the Eleanor Betty Salt Work, Belize. *Research Reports in Belizean* Archaeology 15:275–285.

Feathers, Valerie, Heather McKillop, E. Cory Sills, and Rachel Watson
2017 Excavating the Underwater Shell Deposit at an Ancient Maya Salt Work in Belize: The Eleanor Betty Site. *Research Reports in Belizean Archaeology* 14:289–298.

Fenner, Jack N., and Lori E. Wright
2014 Revisiting the Strontium Contribution of Sea Salt in the Human Diet. *Journal of Archaeological Science* 44: 99–103. doi.org/10.1016/j.jas.2014.01.020.

Ferguson, William M., and Richard E. W. Adams
2001 *Mesoamerica's Ancient Cities.* Rev. ed. University of New Mexico Press, Albuquerque.

Flad, Rowan K.
2007 Rethinking the Context of Production through an Archaeological Study of Ancient Salt Production in the Sichuan Basin, China. In *Rethinking Craft Specialization in Complex Societies: Archaeological Analyses of the Social Meaning of Production*, edited by Zachary X. Hruby and Rowan K. Flad, 108–128. Archeological Papers of the American Anthropological Association 17, Washington DC.

2011 *Salt Production and Social Hierarchy in Ancient China: An Archaeological Investigation of Specialization in China's Three Gorges.* Cambridge University Press, New York.

2013 Salt Technology in Ancient China: Resources and Networks in Two Bronze Age Cases. Paper presented at the XXXV Coloquio de Antropología e Historia Regionales. El Pasado Tecnológico: Cambio y Persistencia, El Colegio de Michoacan, Las Fuentes, Michoacan, Mexico.

Flad, Rowan K., J. Zhu, C. Wang, Pochen Chen, Lothar von Falkenhausen, Z. Sun, and S. Lin
2005 Archaeological and Chemical Evidence for Early Salt Production in China. *Proceedings of the National Academy of Sciences* 102:12618–12622. doi.org/10.1073/pnas.0502985102.

Foias, Antonia
2007 Ritual, Politics, and Pottery Economies in the Classic Maya Southern Lowlands. In *Mesoamerican Ritual Economy*, edited by E. Christian Wells and Karla L. Davis-Salazar, pp. 167–194. University Press of Colorado, Boulder.

Freidel, David
1981 The Political Economies of Residential Dispersion among the Lowland Maya. In *Lowland Maya Settlement Patterns*, edited by Wendy Ashmore, pp. 371–382. University of New Mexico Press, Albuquerque.

Freidel, David A., Marilyn A. Masson, and Michelle Rich
2016 Imagining a Complex Maya Political Economy: Counting Token and Currencies in Image, Text, and the Archaeological Record. *Cambridge Archaeological Journal* 27:29–54. doi.org/10.1017/S0959774316000500.

Freidel, David A., and Jeremy A. Sabloff

1984 *Cozumel: Late Maya Settlement Patterns*. Academic Press, New York.

Freidel, David A., and Justine Shaw

2000 The Lowland Maya Civilization: Historical Consciousness and Environment. In *The Way the Wind Blows: Climate, History, and Human Action*, edited by Roderick J. McIntosh, Joseph A. Tainter, and Susan K. McIntosh, pp. 271–300. Columbia University Press, New York.

Fu, Yonggan

2015 The Salt-Making Workshop Sites of the Shang and Western Zhou Dynasties in Northern Shandong and Their Organization of Production. *Chinese Archaeology* 15:167–174. doi.org/10.1515/char-2015-0017.

Gifford, James C.

1976 *Prehistoric Pottery Analysis and the Ceramics of Barton Ramie and the Belize Valley*. Memoirs of the Peabody Museum of Archaeology and Ethnology 18. Harvard University, Cambridge, Massachusetts.

Gonlin, Nancy

2004 Methods for Understanding Classic Maya Commoners: Structure Function, Energetics, and More. In *Ancient Maya Commoners*, edited by Jon C. Lohse and Fred Valdez Jr., pp. 225–254. University Press of Texas, Austin.

Good, Catharine

1995 Salt Production and Commerce in Guerrero, Mexico. An Ethnographic Contribution to Historical Reconstruction. *Ancient Mesoamerica* 6:1–13. doi.org/10.1017/S0956536100002066.

Gouletquer, Pierre L.

1975 Niger, Country of Salt. In *Salt: The Study of an Ancient Industry*, edited by K. W. de Brisay and K. A. Evans, 47–51. Colchester Archaeological Group, Colchester.

Gouletquer, Pierre L., and Olivier Weller

2015 Techniques of Salt-Making: From China (Yangtze River) to Their World Context. In *Archaeology of Salt: Approaching an Invisible Past*, edited by Robin Brigand and Olivier Weller, pp. 13–28. Sidestone, Leiden.

Graham, Elizabeth

1989 Brief Synthesis of Coastal Site Data from Colson Point, Placencia, and Marco Gonzalez, Belize. In *Coastal Maya Trade*, edited by Heather McKillop and Paul F. Healy, pp. 135–154. Occasional Papers in Anthropology 8, Trent University, Peterborough, Ontario, Canada.

1994 *The Highlands of the Lowlands: Environment and Archaeology in the Stann Creek District, Belize, Central America*. Monographs in World Archaeology 19. Prehistory Press, Washington, DC.

Graham, Elizabeth, and David M. Pendergast

1989 Excavations at the Marco Gonzalez Site, Ambergris Cay, Belize 1986. *Journal of Field Archaeology* 16:1–16. doi.org/10.1179/jfa.1989.16.1.1.

Guderjan, Thomas H.
1988 Maya Maritime Trade at San Juan, Ambergris Caye, Belize. Ph.D. dissertation, Department of Anthropology, Southern Methodist University, Dallas.
Guderjan, Thomas H., and James F. Garber (editors)
1995 *Maya Maritime Trade, Settlement, and Populations on Ambergris Caye, Belize.* Lancaster, Labyrinthos.
Guernsey, Julia
2016 Water, Maize, Salt, and Canoes: An Iconography of Economics at Late Preclassic Izapa. *Latin American Antiquity* 27:340–356. doi.org/10.7183/1045-6635.27.3.340.
Halliwell, Alisha, Andrea Yankowski, and Nigel Chang
2016 Gendered Labor in Pottery and Salt Production in Northeast Thailand. In *Gendered Labor in Specialized Economies*, edited by Sophia E. Kelly and Traci Ardren, pp. 117–157. University Press of Colorado, Boulder.
Hammond, Norman
1972 Obsidian Trade Routes in the Mayan Area. *Science* 178:1092–1093. doi.org/10.1126/science.178.4065.1092.
1975 *Lubaantun: A Classic Maya Realm.* Monographs of the Peabody Museum of Archaeology and Ethnology 2. Harvard University Press, Cambridge, Massachusetts.
1981 Classic Maya Canoes. *International Journal of Nautical Archaeology* 10:173–185.
Hammond, Norman (editor)
1991 *Cuello: An Early Maya Community in Belize.* Cambridge University Press, New York.
Hammond, N., G. Harbottle, and T. Gazard
1976 Neutron activiation and statistical analysis of Maya ceramics and clays from Lubaantun, Belize. *Archaeometry* 18:147–168. doi.org/10.1111/j.1475-4754.1976.tb00157.x.
Harrison, Peter D.
1999 *The Lords of Tikal.* Thames and Hudson, New York.
Harrison-Buck, Eleanor, and Patricia McAnany
2013 Terminal Classic Circular Architecture in the Sibun Valley, Belize. *Ancient Mesoamerica* 24:295–306. doi.org/10.1017/S0956536113000199.
Haviland, William
1985 *Excavations in Small Residential Groups of Tikal, Groups 4F-1 and 4F-2.* Tikal Report 19. University Museum, University of Pennsylvania.
Healy, Paul F., Jaime J. Awe, Gyles Iannone, and Cassandra Bill
1995 Pacbitun (Belize) and ancient Maya Use of Slate. *Antiquity* 69:337–348. doi.org/10.1017/S0003598x00064735.
Healy, Paul F., Heather McKillop, and Bernetta Walsh
1984 Analysis of Obsidian from Moho Cay, Belize: New Evidence on Classic Maya Trade Routes. *Science* 225:414–417. doi.org/10.1126/science.225.4660.414.
Hearth, Nicolas
2012 Organization of Chert Tool Economy during Late and Terminal Classic Peri-

ods at Chan: Preliminary Thoughts Based upon Debitage Analyses. In *Chan: An Ancient Maya Farming Community*, edited by Cynthia Robin, 192–206. University Press of Florida, Gainesville.

Hewitson, Chris

2015 *The Open Pan: The Archaeology and History of the Lion Salt Works*. Research Report 6, Marsten, Lion Salt Works Trust, Cheshire, England.

Hirth, Kenneth G.

1998 The Distributional Approach. *Current Anthropology* 39:451–476. doi.org/10.1086/204759.

2009a Housework and Domestic Craft Production: An Introduction. In *Housework: Craft Production and Domestic Economy in Ancient Mesoamerica*, edited by Kenneth G. Hirth, pp. 1–12. Archeological Papers of the American Anthropological Association, No. 19.

2009b Craft Production, Household Diversification, and Domestic Economy in Prehispanic Mesoamerica. In *Housework: Craft Production and Domestic Economy in Ancient Mesoamerica*, edited by Kenneth G. Hirth, 13–32. Archeological Papers of the American Anthropological Association, No. 19.

Horowitz, Rachel A.

2017 Uneven Lithic Landscapes: Raw Material Procurement and Economic Organization among the Late/Terminal Classic Maya in Western Belize. *Journal of Archaeological Sciences: Reports*. doi.org/10.1016/j.jasrep.2017.01.038.

Houston, Stephen, David A. Stuar, and Karl Taube

2006 *The Memory of Bones: Body, Being, and Experience among the Classic Maya*. University of Texas Press, Austin.

Hutson, Scott, Richard E. Terry, and Bruce H. Dahlin

2017 Marketing within Chunchucmil. In *Ancient Maya Commerce*, edited by Scott Hutson, 241–271. University Press of Colorado, Boulder.

Inomata, Takeshi

2001 The Power and Ideology of Artistic Creation: Elite Craft Specialists in Classic Maya Society. *Current Anthropology* 42:321–349. doi.org/10.1086/320475.

2007 Knowledge and Belief in Artistic Production by Classic Maya Elites. In *Rethinking Craft Specialization in Complex Societies: Archaeological Analyses of the Social Meaning of Production*, edited by Zachary X. Hruby and Rowan K. Flad, pp. 129–141. Archeological Papers of the American Anthropological Association, No. 17. American Anthropological Association, Arlington, Virginia.

Inomata, Takeshi, and Stephen Houston (editors)

2001 *Royal Courts of the Ancient Maya*. 2 vols. Westview Press, Boulder.

Inomata, Takeshi, Daniela Triadan, Erick Ponciano, Estela Pinto, Richard E. Terry, and Markus Eberl

2002 Domestic and Political Lives of the Classic Maya Elites: The Excavation of Rapidly Abandoned Structures at Aguateca, Guatemala. *Latin American Antiquity* 13:305–330. doi.org/10.2307/972113.

Johnston, Kevin J.

2004 The "Invisible" Maya: Minimally Mounded Residential Settlement at Itzan, Peten, Guatemala. *Latin American Antiquity* 15:145–175. doi.org/10.2307/4141552.

Jones, Christopher

2015 The Marketplace at Tikal. In *The Ancient Maya Marketplace*, edited by Eleanor M. King, pp. 67–89. University of Arizona Press, Tucson.

Jones, John G.

2013 Analysis of Sediments from Offshore Locations in Belize, PY30311-1 and PY-30311-10. Manuscript on file with H. McKillop, Department of Geography and Anthropology, Louisiana State University.

Jordan, Jillian M., and Keith M. Prufer

2017 Identifying Domestic Ceramic Production in the Maya Lowlands: A Case Study from Uxbenka, Belize. *Latin American Antiquity* 28:66–87. doi.org/10.1017/laq.2016.3.

Joyce, Rosemary

2018 Honduras in Early Postclassic Mesoamerica. *Ancient Mesoamerica*, pp. 1–14. doi.org/10.1017/S0956536118000196.

Kepecs, Susan M.

1998 Diachronic Ceramic Evidence and Its Social Implications in Northeast Yucatán, Mexico. *Ancient Mesoamerica* 9:121–137.

2003a Chickinchel. In *The Postclassic Mesoamerican World*, edited by Michael E. Smith and Frances F. Berdan, pp. 259–268. University of Utah Press, Salt Lake City.

2003b Salt Sources and Production. In *The Postclassic Mesoamerican World*, edited by Michael E. Smith and Frances F. Berdan, pp. 126–130. University of Utah Press, Salt Lake City.

2005 Mayas, Spaniards, and Salt: Worlds Systems Shifts in Sixteenth-Century Yucatán. In *The Postclassic to Spanish-Era Transition in Mesoamerica: Archaeological Perspectives*, edited by Susan Kepecs and Rani T. Alexander, pp. 117–137. University of New Mexico Press, Albuquerque.

King, Eleanor (editor)

2015 *The Ancient Maya Marketplace: The Archaeology of Transient Space*. University of Arizona Press, Tucson.

Knudson, Kelly J., and Liam Frink

2010 Ethnoarchaeological Analysis of Arctic Fish Processing: Chemical Characterization of Soils on Nelson Island, Alaska. *Journal of Archaeological Science* 37:769–783. doi.org/10.1016/j.jas.2009.11.007.

Kosakowsky, Laura J., Holley Moyes, Mark Robinson, and Barbara Voorhies

2013 Ceramics of Las Cuevas and the Chiquibul: At World's End. *Research Reports in Belizean Archaeology* 10:25–32.

Lentz, David L.

1990 *Acrocomia mexicana*: Palm of the Ancient Mesoamericans. *Journal of Ethnobiology* 10:183–194.

Lentz, David L., Jason Yaeger, Cynthia Robin, and Wendy Ashmore
2005 Pine, Prestige, and Politics of the Late Classic Maya at Xunantunich, Belize. *Antiquity* 79:573–585. doi.org/10.1017/S0003598x00114516.

Li, Shuicheng, and Lothar von Falkenhausen (editors)
2006 *Salt Archaeology in China*, vol. 1: *Ancient Salt Production and Landscape Archaeology in the Upper Yangzi Basin: Preliminary Studies*. Science Publishing, Beijing.
2010 *Salt Archeology in China*, Vol. 2: *Global Comparative Perspectives*. Science Publishing, Beijing, China.

Lin, S. Y.
1940 Salt Manufacture in Hong Kong. *Hong Kong Naturalist* 10(1):34–37.

Locker, A.
2007 In piscibus diversis: The Bone Evidence for Fish Consumption in Roman Britain. *Britannia* 38:141–180.

Lohse, Jon. C., Jaime Awe, Cameron Griffith, Robert M. Rosenswig, and Fred Valdez Jr.
2006 Preceramic Occupations in Belize: Updating the Paleoindian and Archaic Record. *Latin American Antiquity* 17:209–226.

Macintyre, Ian G., M. M. Littler, and D. S. Littler
1995 Holocene History of Tobacco Range, Belize, Central America. *Atoll Research Bulletin* 430, Smithsonian Institution, Washington, DC.

MacKinnon, J. Jefferson, and Susan M. Kepecs
1989 Prehispanic Saltmaking in Belize: New Evidence. *American Antiquity* 53:522–533. doi.org/10.2307/280780.

MacPhail, Richard I, Elizabeth Graham, John Crowther, and Simon Turner
2017 Marco Gonzalez, Ambergris Caye, Belize: A Geoarchaeological Record of Ground Raising Associated with Surface Soil Formation and the Presence of a Dark Earth. *Journal of Archaeological Science* 77:35–51. doi.org/10.1016/j.jas.2016.06.003.

Marcus, Joyce
1983 Lowland Maya Archaeology at the Crossroads. *American Antiquity* 48:454–488. doi.org/10.2307/280556.
1991 Another Pinch of Salt: A Comment on MacKinnon and Kepecs. *American Antiquity* 56:526–527. doi.org/10.2307/280901.
1993 Ancient Maya Political Organization. In *Lowland Maya Archaeology in the Eighth Century A.D.*, edited by Jeremy A. Sabloff and John S. Henderson, pp. 111–183. Dumbarton Oaks Research Library and Collections, Washington, DC.

Martin, Simon, and Nikolai Grube
2000 *Chronicle of the Maya Kings and Queens: Deciphering the Dynasties of the Ancient Maya*. Thames and Hudson, New York.

Martindale Johnson, Lucas R.
2014 Standardized Lithic Technology and Crafting at the "Gateway Group" from Caracol, Belize: Implications for Maya Household Archaeology. *Research Reports in Belizean Archaeology* 11:81–94.

Masatoshi, Kishimoto
2010 The Circulation of Salt in the Ancient Japanese State. In *Salt Archaeology in China*, Vol. 2: *Global Comparative Perspectives*, edited by Shuicheng Li and Lothar von Falkenhausen, pp. 67–135. Science Press, Beijing.

Masson, Marilyn A.
2001 The Economic Organization of Late and Terminal Classic Period Stone Tool Craft Specialist Workshops at Colha, Belize. *Lithic Technology* 26:29–49. doi.org/10.1080/01977261.2001.11720974.
2004 Fauna Exploitation from the Preclassic to the Postclassic Periods at Four Maya Settlements in Northern Belize. In *Maya Zooarchaeology: New Directions in Method and Theory*, edited by Kitty F. Emery, 97–122. Monograph 51, Cotsen Institute of Archaeology, University of California, Los Angeles.

Masson, Marilyn A., and David A. Freidel
2012 An Argument for Classic Maya Era Maya Market Exchange. *Journal of Anthropological Archaeology* 31:455–484. doi.org/10.1016/j.jaa.2012.03.007.

Masson, Marilyn A., Timothy S. Hare, Carlos Peraza Lope, Barbara C. Escamilla Ojeda, Elizabeth H. Paris, Betsy Kohut, Bradley W. Russell, and Wilberth Cruz Alvarado
2016 Household Craft Production in the Prehispanic Urban Setting of Mayapán, Yucatan, Mexico. *Journal of Archaeological Research* 24:229–274. doi.org/10.1007/s10814-016-9091-x.

Masson, Marilyn A., and Shirley B. Mock
2004 Ceramics and Settlement Patterns at Terminal Classic Period Lagoon Sites in Northeastern Belize. In *The Terminal Classic in the Maya Lowlands: Collapse, Transition, and Transformation*, edited by Arthur A. Demarest, Prudence M. Rice, and Don S. Rice, pp. 367–401. University Press of Colorado, Boulder.

McAnany, Patricia A.
2010 *Ancestral Maya Economies in Archaeological Perspective*. Cambridge, Cambridge University Press.

McGee, R. Jon
2002 *Watching Lacandon Lives*. Pearson, New York.

McKee, Karen
2011 Gross Stratigraphy of Paynes Creek and Pelican Cay Cores, Diagram. On file, H. McKillop, Department of Geography and Anthropology, Louisiana State University.

McKillop, Heather
1984 Prehistoric Maya Reliance on Marine Resources: Analysis of a Midden from Moho Cay, Belize. *Journal of Field Archaeology* 11:25–35. doi.org/10.1179/jfa.1984.11.1.25.
1985 Prehistoric Exploitation of the Manatee in the Maya and Circum-Caribbean Areas. *World Archaeology* 16:338–353. doi.org/10.1080/00438243.1985.9979939.
1989 Coastal Maya Trade: Obsidian Densities at Wild Cane Cay, Belize. In *Research in Economic Anthropology*, Supplement 4, edited by Patricia McAnany and Barry Isaac, pp. 17–56, JAI, Greenwich, Connecticut.

1994 Ancient Maya Tree-Cropping: A Viable Subsistence Alternative for the Island Maya. *Ancient Mesoamerica* 5:129–140. doi.org/10.1017/S0956536100001085.

1995 Underwater Archaeology, Salt Production, and Coastal Maya Trade at Stingray Lagoon, Belize. *Latin American Antiquity* 6:214–228. doi.org/10.2307/971673.

1996a Ancient Maya Trading Ports and the Integration of Long-Distance and Regional Economies: Wild Cane Cay in South-Coastal Belize. *Ancient Mesoamerica* 7(1):49–62. doi.org/10.1017/S0956536100001280.

1996b Prehistoric Maya Use of Native Palms: Archaeobotanical and Ethnobiological Evidence. In *The Managed Mosaic: Ancient Maya Agriculture and Resource Use*, edited by Scott L. Fedick, pp. 278–294. University of Utah Press, Salt Lake City.

2002 *Salt: White Gold of the Ancient Maya*. University Press of Florida, Gainesville.

2004 The Classic Maya Trading Port of Moho Cay. In *The Ancient Maya of the Belize Valley*, edited by James F. Garber, pp. 257–272. University Press of Florida, Gainesville.

2005a Finds in Belize Document Late Classic Maya Salt Making and Canoe Transport. *Proceedings of the National Academy of Sciences* 102:5630–5634. doi.org/10.1073/pnas.0408486102.

2005b *In Search of Maya Sea Traders*. Texas A&M University Press, College Station.

2006 *The Ancient Maya*. Norton, New York.

2007a Ancient Mariners on the Belize Coast: Salt, Stingrays, and Seafood. *Belizean Studies* 29(2):15–28.

2007b GIS of the Maya Canoe Paddle Site, K'ak' Naab'. Foundation for the Advancement of Mesoamerican Studies, http://www.famsi.org/reports/05032/index.html.

2009 The Geopolitics of the Coastal Maya Economy in Southern Belize: Relations between the Coastal and Inland Maya. *Research Reports in Belizean Archaeology* 6:55–61.

2010a Underwater Maya: Spatial Analysis of Briquetage and Wooden Buildings at the Paynes Creek Saltworks, Belize, Central America. In *Salt Archaeology in China*, vol. 2: *Global Comparative Perspectives*, edited by Shuicheng Li and Lothar von Falkenhausen, 348–373. Science Press, Beijing.

2010b Ancient Maya Canoe Navigation and Its Implications for Classic to Postclassic Maya Economy and Sea Trade. *Journal of Caribbean Archaeology*, Special Publication 3:93–105.

2011 Rediscovering Stingray Lagoon. *Research Reports in Belizean Archaeology* 8:173–181.

2012 Constructing and Using a GIS for Fieldwork: The Underwater Maya Project. *Research Reports in Belizean Archaeology* 9:267–278.

2015 Evaluating Ancient Maya Salt Production and the Domestic Economy: The Paynes Creek Salt Works and Beyond. *Research Reports in Belizean Archaeology* 12:97–106.

2016a Underwater Maya. Website, http://underwatermaya.com.

2016b Coastal Economies: Comparing Northern and Southern Belize. In *Perspectives on the Ancient Maya of Chetumal Bay*, edited by Debra S. Walker, 279–291. University Press of Florida, Gainesville.

2017a Diving Deeper in Punta Ycacos Lagoon at the Paynes Creek Salt Works, Belize. *Research Reports in Belizean Archaeology* 14:279–288.

2017b Early Maya Navigation and Maritime Connections in Mesoamerica. In *The Sea in History: The Medieval World*, edited by Michel Balard, 701–715. Boydell and Brewer, Bogner Regis, England.

2018 Contingent Multi-Crafting, Surplus Household Production, and the Maya Quest for Salt. *Research Reports in Belizean Archaeology* 15:265–273.

2019 Deterioration of Waterlogged Materials. In *The SAS Encyclopedia of Archaeological Science*, edited by Sandra L. Lopez Varela. Wiley-Blackwell, Malden, Massachusetts.

McKillop, Heather, and Kazuo Aoyama

2018 Salt and Marine Products in the Classic Maya Economy. *Proceedings of the National Academy of Sciences* 115(43):10948–10952. doi.org/10.1073/pnas.1803639115.

McKillop, Heather, George Harlow, April Sievert, Michael Wiemann, and C. Wayne Smith

2019 Demystifying Jadeite: Underwater Maya Discovery at Ek Way Nal, Belize. *Antiquity* 93, in press.

McKillop, Heather, and Paul F. Healy (editors)

1989 *Coastal Maya Trade*. Occasional Paper 8, Department of Anthropology, Trent University, Peterborough, Ontario, Canada.

McKillop, Heather, Linda Howie, and E. Cory Sills

2018 Identifying Salt Consumers by Thin-Section Petrography of Trade Pottery at the Paynes Creek Salt Works, Belize. Unpublished manuscript, on file, H. McKillop, Department of Geography and Anthropology, Louisiana State University.

McKillop, H., L. Jackson, H. Michel, Fred Stross, and F. Asaro

1988 Chemical Source Analysis of Maya Obsidian: New Perspectives from Wild Cane Cay, Belize. In *Archaeometry 88: Proceedings of the Twenty-Sixth International Archaeometry Symposium*, edited by R. M. Farquhar, R.G.V. Hancock, and L. A. Pavlish, pp. 239–244, Department of Physics, University of Toronto, Toronto, Canada.

McKillop, Heather, and John Jones

2018 Ancient Landscape and Sea-Level Rise at Submerged Maya Salt Works, Belize. Unpublished manuscript, on file, H. McKillop, Department of Geography and Anthropology, Louisiana State University.

McKillop, Heather, John Jones, and E. Cory Sills

2008 The Bog People of Belize. Paper presented at the Belize Archaeology Symposium, Belize City.

McKillop, Heather, Aline Magnoni, Rachel Watson, Shannon Ascher, Bryan Tucker, and Terance Winemiller

2004 The Coral Foundations of Coastal Maya Architecture. *Research Reports in Belizean Archaeology* 1:347–358.

McKillop, Heather, and Robin Robertson

2018 Ich'ak'tun: An Early Shell Midden on the South Coast of Belize. Paper presented at the Belize Archaeology Symposium, June 29, San Ignacio, Belize.

McKillop, Heather, and E. Cory Sills

2013a Sustainable Archaeological Tourism of the Underwater Maya Salt Works in Paynes Creek National Park. *Research Reports in Belizean Archaeology* 10:289–296.

2013b Sustainable Archaeological Tourism of the Maya Paynes Creek Project Using 3D Technology. *Anthropology News* May/June: 12–13.

2016 Spatial Patterning of Salt Production and Wooden Buildings Evaluated by Underwater Excavations at Paynes Creek Salt Work 74. *Research Reports in Belizean Archaeology* 13:229–237.

2017 The Paynes Creek Salt Works: A Model for Coastal Belize. In *The Value of Things: Prehistoric to Contemporary Commodities in the Maya Region*, edited by Jennifer Matthews and Tom Guderjan, 67–86. University of Arizona Press, Tucson.

McKillop, Heather, Elizabeth C. Sills, and Vincent Cellucci

2014 The Ancient Maya Canoe Paddle and the Canoe from Paynes Creek National Park, Belize. *Research Reports in Belizean Archaeology* 11:297–306.

McKillop, Heather, E. Cory Sills, and Jessica Harrison

2010a A Late Holocene Record of Caribbean Sea-Level Rise: the K'ak' Naab' Underwater Site Sediment Record, Belize. *ACUA Underwater Archaeology Proceedings* 2010:200–207.

2010b Ancient Vegetation and Landscape of Salt Production in Paynes Creek National Park, Belize. *Research Reports in Belizean Archaeology* 7:245–252.

McKillop, Heather, and Terance Winemiller

2004 Ancient Maya Environment, Settlement, and Diet: Quantitative and GIS Analyses of Mollusca from Frenchman's Cay. In *Maya Zooarchaeology*, edited by Kitty Emery, pp. 57–80, Cotsen Institute of Archaeology, University of California–Los Angeles.

McSweeney, Kendra

1995 The Cohune Palm (*Orbigyna cohune*, ARECACEAE) in Belize: A Survey of Uses. *Economic Botany* 49:162–171. doi.org/10.1007/BF02862920.

Miller, Mary E.

1986 *The Murals of Bonampak*. Princeton University Press, Princeton, New Jersey.

1999 *Maya Art and Architecture*. Thames and Hudson, New York.

Mock, Shirley B.

1994 The Northern River Lagoon Site (NRL): Late to Terminal Classic Maya Settlement, Saltmaking, and Survival on the Northern Belize Coast. Unpublished Ph.D. dissertation, Department of Anthropology, University of Texas, Austin.

Moholy-Nagy, Hattula

1990 The Misidentification of Mesoamerican Lithic Workshops. *Latin American Antiquity* 1:268–279. doi.org/10.2307/972165.

1997 Middens, Construction Fill, and Offerings: Evidence for the Organization of Classic Period Craft Production at Tikal, Guatemala. *Journal of Field Archaeology* 24: 293–313. doi.org/10.2307/530686.

Morehart, Christopher T., David L. Lentz, and Keith M. Prufer

2005 Wood of the Gods: The Ritual Use of Pine (*Pinus* spp.) by the Ancient Lowland Maya. *Latin American Antiquity* 16:255–274. doi.org/10.2307/30042493.

Mosnier, Serge

1994 Le Sel, or blanc des Mayas. *Sciences et avenir* 566:44–49.

Murata, Satoru

2011 Maya Salters, Maya Potters: The Archaeology of Multicrafting on Non-Residential Mounds at Wits Cah Ak'al, Belize. Ph.D. dissertation, Department of Archaeology, Boston University, Boston.

Nance, C. Roger

1992 Guzman Mound: A Late Preclassic Salt Works on the South Coast of Guatemala. *Ancient Mesoamerica* 3:27–46. doi.org/10.1017/S0956536100002273.

Nenquin, J.A.E.

1961 *Salt: A Study in Economic Prehistory*. De Tempel, Gent. Dissertationes Archaeologicae Gandenses, 6.

Nietschmann, Bernard

1973 *Between Land and Water*. Seminar Press, New York.

Nitta, Eijji

1997 Iron-Smelting and Salt-Making Industries in Northeast Thailand. *Indo-Pacific Prehistory Association Bulletin* 3(16): 153–160. http://journals.lib.washington.edu/index.php/BIPPA/article/view/11656/10286.

Olivier, Laurent

2010 The Briquetage de la Seille (Moselle, France): An Iron-Age Proto-Industrial Salt-Extraction Center. In *Salt Archeology in China*, Vol. 2: *Global Comparative Perspectives*, edited by Shuicheng Li and Lothar von Falkenhausen, pp. 239–259. Science Publishing, Beijing, China.

Olivier, Laurent, and J. Kovačik

2006 The "Briquetage de la Seille": Proto-Industrial Salt Production in the European Iron Age. *Antiquity* 80:558–566. doi.org/10.1017/S0003598x00094035.

Parks, Rowland A.

2009 7.1 Quake Rocks Honduras and Belize. *Amadala* (Belize City), May 29. http://amandala.com.bz/news/7-1-quake-rocks-honduras-and-belize/.

Parsons, Jeffrey R.

2001 *The Last Saltmakers of Nexquipayac, Mexico: An Archaeological Ethnography*. Anthropological Paper 92, Museum of Anthropology. University of Michigan, Ann Arbor.

Pendergast, David M.
1969 *The Prehistory of Actun Balam, British Honduras.* Art and Archaeology Occasional Paper 16, Royal Ontario Museum, Toronto.
1970 *A. H. Anderson's Excavations at Rio Frio Cave E, British Honduras (Belize).* Art and Archaeology Occasional Paper 20, Royal Ontario Museum, Toronto.
1971 *Excavations at Eduardo Quiroz Cave, British Honduras (Belize).* Art and Archaeology Occasional Paper 21, Royal Ontario Museum, Toronto.
1974 *Excavations at Actun Polbilche, Belize.* Archaeology Monograph 1. Royal Ontario Museum, Toronto.
1979 *Excavations at Altun Ha*, vol. 1. Royal Ontario Museum, Toronto.
1982 Ancient Maya Mercury. *Science* 217:533–535. doi.org/10.1126/science.217.4559.533.

Pendergast, David M., Elizabeth Graham, Jorge Calvera, and Juan Jardines
2001 Houses in the Sea: Excavation and Preservation at Los Buchillones, Cuba. In *Enduring Records: The Environmental and Cultural Heritage of Wetlands*, edited by Barbara A. Purdy, pp. 71–82. Oxbow Books, Oxford.

Prager, Christian M., and Geoffrey Braswell
2016 Maya Politics and Ritual: An Important New Hieroglyphic Text on a Carved Jade from Belize. *Ancient Mesoamerica* 27(2):267–278. doi.org/10.1017/S095653611600033X.

Prufer, Keith
2005 The Early Classic in Southern Belize: A Regional View from Uxbenka and Ek Xux. *Research Reports in Belizean Archaeology* 2:169–178.

Prufer, Keith M., and W. J. Hurst
2007 Chocolate in the Underworld Space of Death: Cacao Seeds from an Early Classic Mortuary Cave. *Ethnohistory* 54: 273–301. doi.org/10.1215/00141801-2006-063.

Prufer, Keith, Andrew Kindon, and Doug Kennett
2007 Uxbenka and the Founding of Sedentary Communities in Southern Belize. *Research Reports in Belizean Archaeology* 5:241–249.

Prufer, Keith M., Clayton R. Meredith, Asia Alsgaard, Timothy Dennehy, and Douglas Kennett
2016 The Paleoindian Chronology of Tzib Te Yux Rockshelter in the Rio Blanco Valley of Southern Belize. *Research Reports in Belizean Archaeology* 14:321–326.

Prufer, Keith M., Holley Moyes, Brenden J. Culleton, Andrew Kindon, and Douglas Kennett
2011 Uxbenká: The Development of a Complex Polity in the Eastern Periphery of the Maya Lowlands. *Latin American Antiquity* 22:199–223.

Purdy, Barbara (editor)
1981 *Wet Site Archaeology.* Telford Press, Caldwell, New Jersey.
2001 *Enduring Records: The Environmental and Cultural Heritage of Wetlands.* Oxbow Press, Oxford.

Redfield, Robert, and Alfonso Villa Rojas

1962 *Chan Kom: A Maya Village.* University of Chicago Press, Chicago. Originally published 1934, Carnegie Institution of Washington.

Reents-Budet, Dorie

1994 *Painting the Maya Universe: Royal Ceramics of the Classic Period.* Duke University Press, Durham, North Carolina.

2001 Classic Maya Concepts of the Royal Court: An Analysis of Renderings on Pictorial Ceramics. In *Royal Courts of the Ancient Maya*, vol. 1, edited by Takeshi Inomata and Stephen Houston, pp. 195–233. Westview, Boulder, Colorado.

Reents-Budet, Dorie, Ronald L. Bishop, Jennifer T. Taschek, and Joseph W. Ball

2000 Out of the Palace Dumps: Ceramic Production and Use at Buenavista del Cayo. *Ancient Mesoamerica* 11:99–121.

Reese-Taylor, Kathryn

2016 The First Settlers on Chetumal Bay. In *Perspectives on the Ancient Maya of Chetumal Bay*, edited by Debra S. Walker, pp. 33–55. University Press of Florida, Gainesville.

Reina, Ruben E., and John Monaghan

1981 The Ways of the Maya: Salt Production in Sacapulas, Guatemala. *Expedition* 23:13–33.

Reinecke, Andreas

2010 Early Evidence of Salt-Making in Vietnam: Archaeological Finds, Historical Records, and Traditional Methods. In *Salt Archaeology in China*, vol. 2: *Global Comparative Perspectives*, edited by Shuicheng Li and Lothar von Falkenhausen, pp. 137–159. Science Press, Beijing.

Robertson, Robin

2016 Red Wares, Zapatista, Drinking Vessels, Colonists and Exchange at Cerro Maya. In *Perspectives on the Ancient Maya of Chetumal Bay*, edited by Debra S. Walker, pp. 125–148. University Press of Florida, Gainesville.

Robin, Cynthia (editor)

2012 *Chan: An Ancient Maya Farming Community.* University Press of Florida, Gainesville.

Robinson, Mark, and Heather McKillop

2013 Ancient Maya Wood Selection and Forest Exploitation: A View from the Paynes Creek Salt Works, Belize. *Journal of Archaeological Science* 40:3584–3595. doi.org/10.1016/j.jas.2013.04.028.

2014 Fueling the Ancient Maya Salt Industry. *Economic Botany* 68:96–108.

Rochette, Erick T.

2009 Jade in Full: Prehispanic Domestic Production of Wealth Goods in the Middle Motagua Valley, Guatemala. In *Housework: Craft Production and Domestic Economy in Ancient Mesoamerica*, edited by Kenneth G. Hirth, pp. 205–224. Archeological Papers of the American Anthropological Association, No. 19, Washington, DC.

Rosenswig, Robert M.

2015 A Mosaic of Adaptation: The Archaeological Record for Mesoamerica's Archaic

Period. *Journal of Archaeological Research* 23:115–162. https://doi.org/10.1007/s10814-014-9080-x.

Sabloff, Jeremy A.

1975 *Excavations at Seibal, Department of the Peten, Guatemala: Ceramics*. Memoirs of the Peabody Museum of Archaeology and Ethnology 13(2). Peabody Museum of Archaeology and Ethnology, Harvard University, Cambridge, Massachusetts.

Sabloff, Jeremy A., and E. Wyllys Andrews V (editors)

1986 *Late Lowland Maya Civilization: Classic to Postclassic*. University of New Mexico Press, Albuquerque.

Sabloff, Jeremy A., and William L. Rathje (editors)

1975 *A Study of Changing Pre-Columbian Commercial Systems: The 1972–73 Seasons at Cozumel, Mexico*. Monographs of the Peabody Museum of Archaeology and Ethnology, vol. 3. Peabody Museum of Archaeology and Ethnology, Harvard University, Cambridge, Massachusetts.

Saile, Thomas

2010 Earliest Salt-Making in Central Europe: Patterns of Salt Production and Trade in the Neolithic. In *Salt Archaeology in China*, vol. 2: *Global Comparative Perspectives*, edited by Shuicheng Li and Lothar von Falkenhausen, pp. 199–217. Science Press, Beijing.

2015 Competing on Unequal Terms: Saltworks at the Turn of the Christian Era. In *Archaeology of Salt: Approaching an Invisible Past*, edited by Robin Brigand and Olivier Weller, pp. 199–209. Sidestone, Leiden.

Shafer, Harry J., and Thomas R. Hester

1983 Ancient Maya Chert Workshops in Northern Belize. *American Antiquity* 48:519–543. doi.org/10.2307/280559.

Sheets, Payson, Christine Dixon, David Lentz, Rachel Egan, Alexandria Halmbacher, Venicia Slotten, Rocio Herrera, and Celine Lamb

2015 The Sociopolitical Economy of an Ancient Maya Village: Ceren and Its Sacbe. *Latin American Antiquity* 26:341–361. doi.org/10.7183/1045-6635.26.3.341.

Sierra Sosa, Thelma, Andrea Cucina, T. Douglas Price, James H. Burton, and Vera Tiesler

2014 Maya Coastal Production, Exchange, Life Style, and Population Mobility: A View from the Port of Xcambó, Yucatán, Mexico. *Ancient Mesoamerica* 25:221–238. doi.org/10.1017/S0956536114000133.

Sills, Elizabeth Cory

2016 Re-Evaluating the Ancient Maya Salt Works at Placencia Lagoon, Belize. *Mexicon* 38(3):69–74.

2017 Extent of Brine Enrichment for Salt Production at the Ancient Maya Placencia Lagoon Salt Works, Belize. *Research Reports in Belizean Archaeology* 14:271–278.

Sills, E. Cory, and Heather McKillop

2010 The Architecture of Salt Production at the John Spang Site, Paynes Creek National Park, Belize. *Research Reports in Belizean Archaeology* 7:253–258.

2013 Underwater Excavations of Classic Period Salt Works, Paynes Creek National Park, Belize. *Research Reports in Belizean Archaeology* 10:281–288.

2018a Specialized Salt Production during the Ancient Maya Classic Period at Two Paynes Creek Salt Works, Belize: Chan b'i and Atz'aam Na. *Journal of Field Archaeology* 43(6):457–471. https://doi.org/10.1080/00934690.2018.1504543.

2018b Obsidian Source Use at the Paynes Creek Salt Works, Belize. Paper presented at the South Central Conference on Mesoamerica, San Antonio, TX, October 21.

Sills, E. Cory, Heather McKillop, and E. Christian Wells

2016 Chemical Signatures of Ancient Activities at Chan b'i: A Submerged Maya Salt Works, Belize. *Journal of Archaeological Science: Reports* 9:654–662. doi.org/10.1016/j.jasrep.2016.08.024.

Smith, Carol A.

1974 Economies of Marketing Systems: Models for Economic Geography. *Annual Review of Anthropology* 3:167–201.

Smith, C. Wayne

2003 *Archaeological Conservation Using Polymers: Practical Applications for Organic Artifact Stabilization*. Texas A&M University Press, College Station.

Smith, Michael E., and Frances F. Berdan (editors)

2003 *The Postclassic Mesoamerican World*. University of Utah Press, Salt Lake City.

Somers, Bretton Michael

2007 Spatial Analysis of Preserved Wooden Architectural Remains of Eight Late Classic Maya Salt Works in Punta Ycacos Lagoon, Toledo District, Belize. Ph.D. dissertation, Department of Geography and Anthropology, Louisiana State University. Available online at https://digitalcommons.lsu.edu/cgi/viewcontent.cgi?article=4062&context=gradschool_dissertations.

Somers, Bretton, and Heather McKillop

2005 Hidden Landscapes of the Ancient Maya on the South Coast of Belize. *Research Reports in Belizean Archaeology* 2:291–300.

Sordoillet, Dominique, Olivier Weller, Nicolas Rouge, and Martine Buatier

2018 Earliest Salt Working in the World: From Excavation to Microscopy at the Prehistoric Sites of Tolici and Lunca (Romania). *Journal of Archaeological Science* 89:46–55. https://doi.org/10.1016/j.jas.2017.11.003.

Stemp, James, Jaime J. Awe, Keith M. Prufer, and Christophe G. B. Helmke

2016 Design and Function of Lowe and Sawmill Points from the Preceramic Period of Belize. *Latin American Antiquity* 27:279–299. doi.org/10.7183/1045-6635.27.3.279.

Terry, Richard E., Daniel A. Blair, and Eric G. Coronel

2015 Social Chemistry in the Search for Ancient Maya Marketplaces. In *The Ancient Maya Marketplace: The Archaeology of Transient Space*, edited by Eleanor King, 139–167. University of Arizona Press, Tucson.

Thompson, J. Eric S.

1951 Canoes and Navigation of the Maya and Their Neighbours. *Journal of the Royal Anthropological Institute* 79:69–78. doi.org/10.2307/2844504.

Tozzer, Alfred M. (translator)

1941 *Landa's Relaciones de las Cosas de Yucatan*. Papers of the Peabody Museum of Archaeology and Ethnology 18, Harvard University, Cambridge, Massachusetts.

Trik, Aubrey

1963 The Splendid Tomb of Temple 1 at Tikal, Guatemala. *Expedition* 6:2–18.

Valdez, Fred, Jr., and Shirley Mock

1991 Additional Considerations for Prehispanic Saltmaking in Belize. *Latin American Antiquity* 56(3):520–525. doi.org/10.2307/280900.

VandenBosch, Jon C., Lisa J. LeCount, and Jason Yaeger

2010 Integration and Interdependence: The Domestic Chipped Stone Economy of the Xunantunich Polity. In *Classic Maya Provincial Polities: Xunantunich and Its Hinterlands*, edited by Lisa J. LeCount and Jason Yaeger, 272–294. University of Arizona Press, Tucson.

Vogel, Hans Ulrich

2010 From Igneous to Solar Salt: Production Techniques of Maritime Salines in Imperial China. In *Salt Archaeology in China*, vol. 2: *Global Comparative Perspectives*, edited by Shuicheng Li and Lothar von Falkenhausen, 21–41. Science Press, Beijing.

Vogt, Evon Z.

1970 *The Zinacantecos of Mexico: A Modern Maya Way of Life*. Holt, Rinehart, and Winston, New York.

Voorhies, Barbara (editor)

1989 *Ancient Trade and Tribute: Economies of the Soconusco Region of Mesoamerica*. University of Utah Press, Salt Lake City.

Walker, Debra S.

2016 Life and Afterlife at Cerro Maya, Belize. In *Perspectives on the Ancient Maya of Chetumal Bay*, edited by Debra W. Walker, pp. 56–75. University Press of Florida, Gainesville.

Wanyerka, Phil

2005 Epigraphic Evidence of Macro Political Organization in Southern Belize: Evidence From the Early Classic Period. *Research Reports in Belizean Archaeology* 2:179–192.

Watanabe, John M.

2007 Ritual Economy and the Negotiation of Autarky and Interdependence in a Ritual Mode of Production. In *Mesoamerican Ritual Economy: Archaeological and Ethnological Perspectives*, edited by E. Christian Wells and Karla L. Davis-Salazar, pp. 301–322. University Press of Colorado, Boulder.

Watson, Rachel M.

2015 Excavations and Interpretation of Two Ancient Maya Salt-Work Mounds, Paynes Creek National Park, Toledo District, Belize. Ph.D. dissertation, Department of Geography and Anthropology, Louisiana State University, Baton Rouge. Available online at https://digitalcommons.lsu.edu/cgi/viewcontent.cgi?article=2578&context=gradschool_dissertations.

Watson, Rachel, and Heather McKillop
2019 A Filtered Past: Interpreting Salt Production and Trade Models from Two Remnant Brine Enrichment Mounds at the Ancient Maya Paynes Creek Salt Works, Belize. *Journal of Field Archaeology*, in press.

Watson, Rachel, Heather McKillop, and E. Cory Sills
2013 Brine Enriching Slag Heaps or Mounded Remains of Salt Makers Homes? Earthen Mounds in the Mangroves at the Paynes Creek Salt Works. *Research Reports in Belizean Archaeology* 10:297–304.

Wauchope, Robert
1938 *Modern Maya Houses: A Study of Their Archaeological Significance*. Carnegie Institution of Washington Publication 504, Washington, DC.

West, Robert C., and John P. Augelli
1989 *Middle America: Its Land and Peoples*, 3rd ed. Prentice Hall, Englewood Cliffs, New Jersey.

Whittaker, John C., Kathryn A. Kamp, Anabel Ford, Rafael Guerra, Peter Brands, Jose Guerra, Kim McLean, Alex Woods, Melissa Badillo, Jennifer Thornton, and Zerifeh Eiley
2009 Lithic Industry in a Maya Center: An Axe Workshop at El Pilar, Belize. *Latin American Antiquity* 20:134–156. doi.org/10.1017/S1045663500002558.

Wilk, Richard
1983 Little House in the Jungle: The Causes of Variation in House Size Among Modern Maya. *Journal of Anthropological Archaeology* 2:99–116. doi.org/10.1016/0278-4165(83)90009-0.
1987 The Ketchi and the Settlement of the Toledo District. *Belizean Studies* 15(3):33–50.
1988 Maya Household Organization: Evidence and Analogies. In *Household and Community in the Mesoamerican Past*, edited by Richard R. Wilk and Wendy Ashmore, pp. 135–151. University of New Mexico Press, Albuquerque.

Williams, Eduardo
1999 The Ethnoarchaeology of Salt Production at Lake Cuitzeo, Michoacan, Mexico. *Latin American Antiquity* 10:400–414. doi.org/10.2307/971964.
2003 *La Sal de La Tierra*. El Colegio de Michoacan, Zamora, Michoacan.

Wing, Elizabeth
1975 Animal Remains from Lubaantun. In *Lubaantun: A Classic Maya Realm*, by Norman Hammond, pp. 379–383. Peabody Museum Monograph 2, Harvard University, Cambridge, Massachusetts.

Woodfill, Brent K. S., Brian Dervin Dillon, Marc Wolf, Carlos Avendaño, and Ronald Canter
2015 Salinas de Los Nueve Cerros, Guatemala: A Major Economic Center in the Southern Maya Lowlands. *Latin American Antiquity* 26:162–179. doi.org/10.7183/1045-6635.26.2.162.

Wright, A. Charles S., R. H. Romney, R. H. Arbuckle, and V. E. Vial
1959 *Land in British Honduras: Report of the British Honduras Land Use Survey Team*. Colonial Research Publication 24, Colonial Office, London.

Wright, Lori

2005 Identifying Immigrants to Tikal, Guatemala: Defining local Variability in Strontium Isotope Ratios of Human Tooth Enamel. *Journal of Archaeological Science* 32:555–566. doi.org/10.1016/j.jas.2004.11.011.

Yankowski, Andrea

2008 Earthenware Production and Trade: Using Ethnographic Data and Petrographic Analyses to Compare Prehistoric and Contemporary Pottery Traditions from the Island of Bohol, Philippines. *Bulletin of the Indo-Pacific Prehistory Association* 28:1–5.

2010 Traditional Technologies and Ancient Commodities: An Ethnoarchaeological Study of Salt Manufacturing and Pottery Production in Bohol, Central Philippines. In *Salt Archaeology in China*, vol. 2: *Global Comparative Perspectives*, edited by Shuicheng Li and Lothar von Falkenhausen, pp. 161–181. Science Press, Beijing.

Yankowski, Andrea, and Puangthip Kerdsap

2013 Salt-Making in Northeast Thailand: An Ethnoarchaeological Study in Tambon Phan Song Khram, Nakhon Rushmasima Province, Northeast Thailand. *Silpakorn University Journal of Social Sciences, Humanities, and the Arts* 13(1):231–252.

Yankowski, Andrea, Puangthip Kerdsap, and Nigel Chang

2015 Please Pass the Salt: An Ethnoarchaeological Study of Salt and Salt-Fermented Fish Production, Use, and Trade in Northeast Thailand. *Journal of Indo-Pacific Archaeology* 37:4–13. doi.org/10.7152/jipa.v37i0.14711.

INDEX

Page numbers in *italics* refer to illustrations.

HEATHER MCKILLOP is Thomas and Lillian Landrum Alumni Professor in the Department of Geography and Anthropology at Louisiana State University. She specializes in Maya archaeology and has authored several books, including *The Ancient Maya*, *In Search of Maya Sea Traders*, and *Salt: White Gold of the Ancient Maya*.

MAYA STUDIES

Edited by Diane Z. Chase and Arlen F. Chase

Salt: White Gold of the Ancient Maya, by Heather McKillop (2002)
Archaeology and Ethnohistory of Iximché, by C. Roger Nance, Stephen L. Whittington, and Barbara E. Borg (2003)
The Ancient Maya of the Belize Valley: Half a Century of Archaeological Research, edited by James F. Garber (2003; first paperback edition, 2011)
Unconquered Lacandon Maya: Ethnohistory and Archaeology of the Indigenous Culture Change, by Joel W. Palka (2005)
Chocolate in Mesoamerica: A Cultural History of Cacao, edited by Cameron L. McNeil (2006; first paperback printing, 2009)
Maya Christians and Their Churches in Sixteenth-Century Belize, by Elizabeth Graham (2011)
Chan: An Ancient Maya Farming Community, edited by Cynthia Robin (2012; first paperback edition, 2013)
Motul de San José: Politics, History, and Economy in a Maya Polity, edited by Antonia E. Foias and Kitty F. Emery (2012; first paperback edition, 2015)
Ancient Maya Pottery: Classification, Analysis, and Interpretation, edited by James John Aimers (2013; first paperback edition, 2014)
Ancient Maya Political Dynamics, by Antonia E. Foias (2013; first paperback edition, 2014)
Ritual, Violence, and the Fall of the Classic Maya Kings, edited by Gyles Iannone, Brett A. Houk, Sonja A. Schwake (2016)
Perspectives on the Ancient Maya of Chetumal Bay, edited by Debra S. Walker (2016)
Maya E Groups: Calendars, Astronomy, and Urbanism in the Early Lowlands, edited by David A. Freidel, Arlen F. Chase, Anne S. Dowd, and Jerry Murdock (2017)
War Owl Falling: Innovation, Creativity, and Culture Change in Ancient Maya Society, by Markus Eberl (2017)
Pathways to Complexity: A View from the Maya Lowlands, edited by M. Kathryn Brown and George J. Bey III (2018)
Water, Cacao, and the Early Maya of Chocolá, by Jonathan Kaplan and Federico Paredes Umaña (2018)
Maya Salt Works, by Heather McKillop (2019)

CPSIA information can be obtained
at www.ICGtesting.com
Printed in the USA
BVHW040223160419
545599BV00005B/9/P

9 780813 056333